THE VITAL LANDSCAPE

THIS BOOK IS DEDICATED TO MY PARENTS AND GRANDPARENTS, PARTICULARLY
TO THE MEMORY OF MAX LAMERS

The Vital Landscape

Nature and the Built Environment in Nineteenth-Century Britain

William M. Taylor

Routledge
Taylor & Francis Group

LONDON AND NEW YORK

First published 2004 by Ashgate Publishing

Reissued 2018 by Routledge
2 Park Square, Milton Park, Abingdon, Oxon OX14 4RN
605 Third Avenue, New York, NY 10017

First issued in paperback 2021

Routledge is an imprint of the Taylor & Francis Group, an informa business

Typeset in Palatino by Bournemouth Colour Press, Parkstone

A Library of Congress record exists under LC control number: 2003070816

Notice:
Product or corporate names may be trademarks or registered trademarks, and are
used only for identification and explanation without intent to infringe.

Publisher's Note
The publisher has gone to great lengths to ensure the quality of this reprint but
points out that some imperfections in the original copies may be apparent.

Disclaimer
The publisher has made every effort to trace copyright holders and welcomes
correspondence from those they have been unable to contact.

ISBN 13: 978-0-815-39839-4 (hbk)
ISBN 13: 978-1-351-14480-3 (ebk)
ISBN 13: 978-1-138-35769-3 (pbk)

DOI: 10.4324/9781351144803

Contents

List of figures

Where unacknowledged, photographs and figures are by the author or from his collection.

Acknowledgements

In my efforts I am much indebted to scholars from a number of fields. Gary Wickham and Geoffrey London offered much-needed advice and support in the early stages of my research. Paul Hirst, David McNeil and Grahame Thompson offered comments on my initial thesis on biology, functionalism and glasshouses, while Michael Levine came to my rescue on numerous occasions, helping me pull diverse thoughts and various words together. For other forms of assistance, criticism and advice I am grateful to Greer Bradbury, Annette Condello, John Dixon-Hunt, John Macarthur and Nicole Sully. Library staff at the University of Western Australia were always helpful. Leslie Price and Kate Pickard at the Royal Botanic Gardens provided research material and a warm welcome to the archives at Kew.

Sections of this book were adapted from my previously published work. Discussion of the Palm House appeared in two articles in *Studies in the History of Gardens & Designed Landscapes*, formerly *The Journal of Garden History*, volumes 15:4 and 18:2 (Winter 1995 and Summer 1998 respectively). The garden cemetery was discussed in *National Identities*, vol. 1:1 (1999) and Robert Kerr's book *The Gentleman's House* in the journal *Design Issues*, vol. 18:3 (Summer 2002). I am grateful to the respective publishers for permission to include parts of these articles here. Research for this project was made possible by a grant from the Australian Research Council. Publication was made possible partly by the support of The Australian Academy of the Humanities.

Frontispiece
Upper gallery of
the Palm House,
Royal Botanic
Gardens at Kew,
by Decimus
Burton and
Richard Turner,
1844–48

Preface

This work develops thoughts on architecture and landscape architecture. It grew from an interest in glasshouses, though the issues raised in the following chapters are not contained or explained by any one type of building or landscape alone.

It seems a particularly challenging task today to describe what architects and landscape architects 'do' in terms of what they 'know'. Obviously, designs are not only the outcome of ideas, but constraints of practice as well. Particular challenges arise from functional considerations, new materials and ways of using them, along with a desire for inventiveness and novel forms. Practical constraints of a different order come into play arising from various social, political and economic concerns. An appreciation of developments in science and their impact on natural and built environments past and present and the quickening pace of technological change introduce further flux into the mix of ideas and practices facing students and professionals, theorists and historians. Recent developments in ecology and evolutionary science lead one to question the common distinction between natural and human-made spheres of habitation as the organic and inorganic worlds appear even more closely interrelated and subject to manipulation than ever before. In different ways novel developments like the Eden Project in Cornwall (2001), fields like genetics and evolutionary psychology, circumstances of climate change and diminishing natural resources – even sick-building syndrome – push the boundary between these two domains of human existence.

One can begin to unpack these issues by considering the really big ideas that bring together concepts from various fields of inquiry and the material contexts they arise from. The 'environment' is itself an immensely important idea. Human understanding of nature has obviously changed throughout the history of Western thought. However varied its forms and forces may have appeared in the past, though, there persists a longstanding view of the distinctiveness of our habitation of the earth. This view clearly prompts much environmental thinking. The idea of habitable space or the figure of the 'inhabitant' of both natural and built environments is a second important, though easily overlooked, focus of theory and practice, history and present-day concerns. This figure entails a view of human beings as uniquely suited to their surroundings as members of other species are to theirs. Implicit in the notion of professional practice is the expectation that designers create unique, healthy and pleasing environs for each and every client. This raises a third big idea, namely the practice of 'design' itself. What happens at the drawing board today is different to what architects and 'landscape gardeners' did two hundred, much less five hundred, years ago. One could even argue that ordinary people occupied space differently in earlier times as well. There was hardly an 'environment' to speak about in any modern sense of the word. There was hardly guidance to design for the numerous relationships arising between 'man, his culture, and his living and non-living surroundings' – the subject of environmental studies today.[1] All things considered, one of the best ways to reflect on architecture and landscape architecture is to think about the history of the big ideas that impinge on these disciplines.

In view of such observations glasshouses may seem of little consequence – at first. My interest in these structures grew from a particular study of the curvilinear iron-and-glass buildings devised or otherwise promoted by John Claudius Loudon (1783–1843), the landscape gardener, architectural critic and social reformer whose ideas were to shape nineteenth-century attitudes to the house and garden. His ideas were derived from Scottish science and Enlightenment, and a noticeable moral fervour permeates his written works.[2] Melanie Simo has observed a certain parallel between Loudon's ideas and those of Jeremy Bentham (1748–1832).[3] Whereas the great utilitarian philosopher sought to find in nature laws to govern human society, Loudon was concerned to adapt building form to the order of the natural world. The point of such a comparison is not to assert that one was a 'naturalist' and the other an 'environmentalist' in the narrow sense of either term; rather, what is interesting is that, along with many of their contemporaries, Bentham and Loudon turned an age-old perception of the purposeful arrangement of creation to practical ends. This entailed the cultivation of praiseworthy sentiments and values and desirable forms of behaviour. Loudon was able to participate in such a novel undertaking, or at least try to relate form to particularities of context, because attitudes to nature had changed markedly, certainly over the course of his lifetime. This is why his writings stand out among other references in this book.

Conceived of and built between 1815 and 1831, Loudon's curvilinear glasshouses were a response to Sir George Mackenzie's advice to horticulturalists to 'Make the surface of your green-house parallel to the vaulted surface of the heavens or to the plane of the sun's orbit.'[4] In a lecture to the Horticultural Society of London in 1815, Mackenzie suggested that the most suitable shape for such buildings was hemispherical, so that light would strike vertically both in the summer and when the sun was low in the winter. He proposed that its ground plan be semi-circular and face south, so that the resulting quarter sphere would receive morning, noon and evening sun while the provision of a masonry wall to the north would more easily retain the sun's heat. The purpose of these measures was to effectively lengthen growing seasons, facilitate the introduction of new plant species and improve the quality of domestic produce. More generally these efforts reflect a desire to adopt a more reasoned approach to constructions of this sort. Mackenzie's proposals led to a number of scientific improvements to the glasshouse, a building type that was to change dramatically in both form and meaning as the century unfolded.

There is but one of these buildings still standing today that bears testimony to Mackenzie's curvilinear principles. Built sometime between 1820 and 1825 by the manufacturing firm of W. & D. Bailey with which Loudon collaborated for a number of years, it stands centre stage at Bicton Gardens near Budleigh Salterton, Devon (see Figure p. XVI). A more enduring legacy is the outcome of the collaboration of English architect Decimus Burton and Irish engineer Richard Turner, both of whom turned to science to devise a novel and 'practical' building form. This is the great Palm House at the Royal Botanic Gardens at Kew, built in the years 1844 to 1848 and restored at various times, most notably on the occasion of its complete reconstruction in 1988 (see Frontispiece).[5] The Palm House is arguably the most famous of iron-and-glass buildings, celebrated as an icon of Victorian ingenuity and technological prowess. It enlarged the boundaries of scientific enterprise to encompass the domain of culture as well as nature for two reasons. First, it was a site for debates over architectural style, form and function. Second, it was one of a number of large structures built in the 1830s and 1840s that introduced members of the public to novel environments and prospects for improving more familiar ones.

The glasshouse helped the British public envisage alternative worlds and ways of living. This is how it is valued in this book and why it forms a starting point and reference for other kinds of buildings and landscapes that were similarly provocative. The architectural historian Nikolaus Pevsner (1902–83) portrayed Loudon's curvilinear structures as forerunners of modern, functional design. They were as 'assertive a profession of faith in iron' as the market and exhibition halls, railway stations and bridges made possible by the tremendous growth of British cities in the early nineteenth century. They reveal much about the fast-growing exchange of materials and goods that made the modern metropolis possible.[6] The Palm House at Kew was completed three years before Joseph Paxton built the Crystal Palace for the Great Exhibition of 1851, a building which is commonly regarded as the culmination of, and catalyst for, new ideas regarding society, the built and natural environments and the impact of science and technology on all three.[7]

Like the rattan palms in Kew's conservatories, the issues raised by my appreciation of these structures threatened to exceed the study of plants and glasshouses, iron and glass alone. Initially my interests in glasshouses served to identify one source for contemporary design theory in the concerns of nineteenth-century horticulture and botanical science. It seemed at the time that the theory and practice of 'function' and 'functionalism' formed a link between an ensemble of emerging biological sciences and efforts to design buildings and landscapes with greater purpose, psychological effect and moral intention. It seemed that whereas biologists value forms of organic life

The Palm House, Bicton Gardens, Devon, built by manufacturers W. and D. Bailey according to designs by John Loudon, 1820–25

and the evolution and adaptation of species to unique geographic regions and climates, writers on architecture and gardening in the nineteenth century commonly encouraged the comprehensive design of buildings and landscapes to suit unique sites, uses and occupants. These ideas were not specific to Britain alone. France and Germany were equally important sources of functionalist thinking at various times. Developments in science and historical circumstances in Britain, though, formed a particularly favourable context for exploring the order of things, in both the natural and human-made worlds.

My initial interest in botany and glasshouses was expanded through further reading in philosophy and the history of science, politics and social theory, ecology and landscape studies, as well as architectural history. It soon became obvious that as the nineteenth century unfolded, references to the functional and the organic were more than simply a set of descriptive terms borrowed from biology. They served broader purposes than the needs of architects in the design of specialist buildings such as glasshouses and crystal palaces. For the theorist and designer, the professional and lay-person alike, references to the function of things and the organic wholeness of living spaces point to an entirely new set of issues and objects of concern. Foremost among these were the environment, the inhabitant and the idea of design. While these concerns were prefigured by studies of plant life in the closed world of the conservatory, they encouraged new modes of environmental awareness in other spheres of human interest such as the Victorian house and garden.

The environment is an idea that owes much to developments in biology and related disciplines like botany and zoology, ecology and evolutionary science in the nineteenth century. It was to have a particularly profound impact on the 'self-awareness' of the inhabitants of public and private spaces – those inside buildings and outdoors, in glasshouses and more familiar, everyday places. The 'sciences of life', as these disciplines are commonly called, recast relations between human beings and other species and the world of non-living things. Life science reinvigorated a longstanding belief in the unique identity of humankind in some ways and challenged it in others. In this regard, the legacy of Charles Darwin is clear. Awareness of the vital processes binding all living creatures to the environment reinforced a sense of personal integrity and wholesomeness based on one's observance of, and obedience to, the order of nature. Environmental awareness on the part of the residents of British cities helped them adapt to the circumstances of their life and its surroundings. Awareness of various manners of adaptation and the distinctiveness of human behaviours, societies and cultures cast 'character' as an important marker of individuality.

The Vital Landscape aims to show how environmental awareness was the result of efforts to accommodate nature, to make room, alongside people, for its forms and processes, species and inanimate matter through works of architecture and landscape gardening in nineteenth-century Britain. In making room for nature, places like glasshouses and the Victorian house and garden revealed its novelty. These built spaces turned thoughts to the situation of humankind in the natural world and raised concerns for the identity, integrity and character of residents. In turn, these concerns helped form the disciplines of architecture and landscape architecture and familiar design practices. This book seeks to show that the built environment has been a means of raising awareness of both the history and the significance of our occupation of the earth along with requirements for sustenance, warmth and light. It argues that the art of designing spaces for shelter and meaningful habitation is a reflection of how we position ourselves in relation to nature, other living creatures and non-living things. Architecture and landscape theorists would do well to attend to the influence of the life sciences on the genesis of their disciplines.

Introduction

By the middle of the nineteenth century it was as though *all* buildings were made of glass. It seems they were either transparent for practical purposes as places of human habitation *or* mirrors of society, reflecting the desires of residents for novelty, style and fashion. This is the complex view of the built environment that comes across in period books on architecture and landscape gardening. It is telling that in many of these books references and terms are drawn from science, notably the sciences of life. In domestic treatises and encyclopaedias in Britain in particular, houses and gardens are described in terms of their formal 'character'. One finds the common insistence that rooms and passageways be arranged to suit their 'function'. Designs appear to suit residents of various 'types', with creative responsibilities distributed among architects, engineers and landscape gardeners forming the professional 'classes'.

These terms, derived from eighteenth-century natural history, persisted in an age of biology and are particularly revealing. They suggest that in the nineteenth century the study of natural history and biology was valued not merely for what nature revealed about living creatures, plants and animals per se. Rather, the study of nature revealed the meanings that the world of living things held for humans. Humankind shared the world with many creatures, but they alone possessed capabilities for self-reflection, imagination and invention, or so it was commonly assumed. They were set apart from the rest of creation by living in communities with distinctive cultures. In the nineteenth century references to the forms and purposes of things were not merely descriptive. They became a point of correspondence between the concerns of architects, horticulturalists and landscape gardeners for the growing reality of organic existence on the one hand and the influence of new social formations and powers on the other. These references prefigure contemporary concerns and practices.

The order of things

The historical context for these developments and the setting for this book are best related by describing two distinct pairs of circumstances. The first pair –

both sequences of events – ended in deaths of differing sorts, though neither caused much grief. Consider the reflections on gardening in 1654 by Sir Hugh Platt in a treatise published in several editions during the mid-seventeenth century and titled *The Garden of Eden*. It was described as a work of natural history, a detailed description of the flowers and fruits growing in England with particular advice on 'how to advance their nature and growth'. In his book Platt defends the unusual horticultural practices of George Ripley (1415–90), describing how the ill-fated alchemist and author of 'hermetico-poetical' works 'suffered death ... for making a pear tree to fructify in winter' (see Figure 0.1). To this curious revelation the author hastens to add that it was not the crime of conjuration, but the denial of Ripley's 'medicine' – presumably a violation of the protocols of alchemy – that killed him.[1]

Next consider a different kind of demise – rather more widespread – involving the unfortunate state of affairs in the Palm House at Kew Gardens, in August 1867. The date was just 18 years after the building was completed by Decimus Burton and Richard Turner and over 50 since George Mackenzie envisioned its curvilinear form (see Figure 0.2). In a letter addressed to his superiors by Joseph Dalton Hooker, administrator of the Gardens at the time and son of Sir William Hooker, its first director, we discover a scene worth relating at length where:

evil is progressive. As the plants have died, or been removed to warmer houses, or have lost their foliage, so is the heat and moisture that these intercepted now dispersed; as the climbers on the rafters have been successively killed, so has the radiation of cold from the iron and glass increased, and as the soil has parted with its heat. ... The plants have made no growth, and very few and small leaves this summer. The leaves are already turning yellow, and are so sparse that the pots and tubs are everywhere seen, and a person standing outside the north wing can see the Museum building through it all along, that is to say, through sixteen parallel rows of plants. Most of the climbers were killed down last winter, and we have not been able this summer to raise them a quarter of the way up the rafters and we have thus no screen from the sun in summer or cold in winter ... I have removed all the most tender plants into the other houses, which we have crowded to excess, and have filled up with about 1500 greenhouse plants instead for appearance's sake; till now there is not a single young palm on the shelves, and the term Palmhouse is a misnomer.[2]

With hindsight, the circumstances accompanying Ripley's death seem strange, even amusing. In *The Garden of Eden* the alchemist's demise appears to have been caused by one of two things. It perhaps followed his infringement of natural law whereby the order of things, plant, animal and humankind – their forms, arrangement and manner of existence – was fixed and immutable. Or, it was caused by his failure to exercise sufficient skill in practising his medicine or craft. In other words, Ripley died because he did not understand nature well enough. In either case, it is not simply the prospect of forcing pear trees to bear fruit in winter that seems extraordinary. Rather, what seems to us so curious today is the means employed to do so. Alchemy is generally considered an outdated practice to say the least.[3]

By comparison, events leading to the loss of the botanical collection in the Palm House centuries later seem disquietingly familiar. Unlike the alchemist,

8 *ryrus hyemalis.*
The Winter Peare tree.

0.1 'The Winter Peare tree'; in John Gerard, *The Herball or Generall Historie of Plantes* (1633). Gerard's book was primarily a natural history rather than an alchemical work, though it included, along with recognizable species, plants and animals that came to be regarded as mythical

0.2 View of the Palm House and reflecting pond, Royal Botanic Gardens at Kew, by Decimus Burton and Richard Turner, 1844–48

we have come to understand death and disease in causal terms. Both are likely to be the function of some deficiency of nature or nurture, even of 'environment'. We rely on a range of technological remedies to forestall the inevitability of decay. Joseph Hooker's search for an explanation for this calamity is understandable, though we, like him, are uncertain as to its precise cause. Perhaps, as accounts variously suggest, it was a function of the structure's green-tinted glass or a faulty heating system.[4] Perhaps it was the delayed effect of a particularly bitter, smoke-ridden London winter. After all, the structure was completed in a century with extreme weather that not only deprived the palms of 'the greatest possible quantity of rays from the sun' so desired by George Mackenzie, but threatened to sap the vitality of metropolitan residents as well.

The preceding circumstances beg a particular orientation to history and the sense that can be made of it. Both events are worth considering for what they reveal about changing ideas and attitudes to nature and our dependence on it. Platt's account of the unfortunate alchemist's fate allows us to appreciate *other* ways of thinking and perhaps to feel secure in the certainties of our own brand of modern, scientific knowledge as opposed to Ripley's outmoded beliefs.[5] This is the weight the episode is made to bear in the early pages of John Hix's key reference on iron-and-glass architecture, *The Glasshouse*.[6] In histories of the Palm House at Kew, the events of 1867 are an interesting aberration. Authors grapple with the past to portray the development of a unique building form as the rational application of science to horticultural and technical problems. Humans usually come across well in such accounts. They appear driven by some universal need or perhaps, as the authors of one book assert, by our 'perception of the needs of tender plants' or our 'desire to enjoy them'.[7]

The substance of Hooker's letter invites us to revisit the Palm House at Kew Gardens and look beyond its status as an iconic work of architecture in an industrial age. The scene it describes makes us appreciate that the world contained within the Victorian glasshouse was not simply one of rare species and exotic beauty. While such novel buildings formed a space in which awareness of the fullness of nature and functional design came together, the great conservatory at Kew, like its curvilinear antecedents and buildings it subsequently inspired, is compelling for other reasons. Perhaps the Palm House proves most interesting not for *being* inherently functional, but for not *working* at all, at least not as its architect and engineer and their clients had originally intended. It is perhaps neither the quantity nor clarity of light admitted by the curvilinear form of the Palm House that should be studied more closely, nor the spectacle of exotics within. Rather, some lingering obscurity, some troubling absence or unanswered question caused the leaves to yellow and vines to wither. Maybe this relatively undervalued detail of glasshouse history might illuminate our complex relationship to buildings and the landscapes they accentuate today.

Regardless of how one views the past, one thing seems obvious given the sequence of events related so far. That is, in the nearly four hundred years separating the fate of Ripley and Kew's prized specimens, it became not only possible, but acceptable, to manipulate the laws of nature. It was worth creating 'artificial climates', as they came to be known. Moreover, a fundamental link was forged between the necessity of preserving organic life and an understanding of the function played by earth, air and water in keeping things alive, other living beings as well as ourselves. The alchemist's insight into the divinely inspired order of creation was lost, but it was replaced by the notion of environmental awareness. This idea is invoked, though rarely explored, in accounts of nineteenth-century architecture and landscape design.[8]

The appearance and purpose of things

If the first pair of historical events raises opposing ideas about the order of nature, the second entails the outward 'show' of order. While the Director of Kew Gardens may have been unsure of the exact cause of his collection's lack of vitality, he was more certain just what the interior of a proper palm house *should* look like. He was confident enough to move in additional plants to compensate for the rapidly wilting tropical scenery, all for 'appearance's sake'. Remaining with plants and glasshouses, iron and glass for a while longer, the second pair of circumstances queried here is less dramatic than the previous one, though equally illustrative of broader concerns.

First, in promoting Mackenzie's ideas about curvilinear conservatories John Loudon urged his readers to seek advice when planning a glasshouse. This was by far preferable to the routine of hiring a tradesman whose ideas on the subject were in no way 'matured', or worse, hiring architects who generally displayed little 'taste' in their designs for this novel building form. The year in which *Remarks on the Construction of Hothouses* (1817) appeared is equally notable for the construction of the Royal Pavilion in Brighton. Loudon condemned this exotic confection of a building for its load-bearing columns falsely covered with metallic palm fronds in a shameful 'mimicry of nature'[9] (see Figure 0.3). His dismissal of its celebrated architect, John Nash, betrays a belief commonly expressed in treatises of the period that buildings be arranged for reasons other than mere style or caprice.[10] Rather, true architecture required expert knowledge whereby the designer assumed responsibility for a building's honesty. Like other nineteenth-century commentators, notably John Ruskin (1819–1900), Loudon invested ethical meaning in form where the integrity of the designer, their adherence to a moral code or the wholeness or even wholesomeness of their character, was mirrored in their plans. It was reflected in a certain honest or straightforward disposition of forms, their unpretentiousness or structural soundness. These ideas were not expressed solely by nineteenth-century commentators, though they acquired greater urgency at that time when it was obvious there were so many ways to build.

Second, a related concern to discern meaning in form is evident in attempts to identify the true designer of Kew's Palm House. These entail a search for authorship common in historical research and particularly evident in histories of novel nineteenth-century building forms in which the roles of architect and engineer, designer and manufacturer are not so easily distinguished.[11] Historians have compiled an impressive body of archival evidence, correspondence, drawings and specifications in their efforts to reveal the often contradictory ideas expressed by early proposals for the Palm House. Burton, an architect with a mind for science, proposed a curvilinear form for the building. Turner, a 'practical' character with an eye for economy of plan and structure, proposed a form similar in some ways though clothed in a style Burton considered out of place. Both men offered reasoned explanations to support their schemes.

What sense can be made of Loudon's 'call' for reasonable building form and the historian's 'use' of form to reveal the hand responsible for designing a glasshouse? What, in other words, connects these efforts? Partly, it is a moral imperative associated with the broad aim to adapt built form to the way the natural world operates as evidenced, say, in a functionally designed glasshouse. It follows the expectation very much in evidence in nineteenth-century thought that form, whether living or human-made, should not only *appear* ordered, but should express or otherwise manifest some essential function or purpose. Given that nature's arrangements were organic, immediate and transparent, so too should be human constructions, lest the latter be denounced as trickery. Likewise, the historian – and we too, when looking back on the past – expect to find some reason, purpose or rationale guiding the evolution of the built environment, however indistinct or varied it may appear. Basically, this is because *we* understand what adaptation means. We like to consider buildings, past and present, to somehow serve and represent an adaptive process whereby their occupants made themselves at home in their surroundings.

0.3   The Royal
Pavilion,
Brighton, John
Nash, 1817–21

Connecting the way we appreciate buildings and landscapes with the way
our nineteenth-century predecessors did is not only a moral imperative, but
also a common language for expressing it. It is a language of forms and
functions, appearances and purposes, means and ends, causes and effects –
terms to derive a sense of the spaces one inhabits as well as to describe what
they looked like.[12] Seen from a broad perspective, it is evident in a number of
historical instances where individuals sought meaning in their surroundings.
It was evinced by some natural philosophers in the eighteenth century, for
instance, as they described the function of a particular animal limb in terms of
its resemblance to a mechanical device or lever. The appearance of specimens
in the glass spheres and retorts used in experiments allowed others to deduce
the cause of a plant's reaction to light or heat. In a different context, Victorian
homemakers were encouraged to deduce the effect of a particular room on
family health from its appearance, the presence or absence of dusty, begrimed
surfaces or smoky atmospheres. Given the popularization of science,
particularly in the second half of the nineteenth century, the language of
forms, functions and appearances was a means whereby knowledge of the
necessities of domestic life was acquired by ordinary people.

Analogues of nature

If the circumstances of Ripley's death and the scene in the Palm House raise
a context for thinking about nature, the expectation of purposeful building

form in the past and present reveals a common ground for decision-making. While the emergence of environmental awareness provides the setting for the buildings and landscapes discussed in the following chapters, the idea of the environment foregrounds the debates that enlivened their design, the technologies that built them and the uncertainties that plagued their residents. The task at hand is to describe what their designers *did* in the past in terms of what they *knew*. In this regard the built environment serves as both a looking glass onto history, its events and contingencies, and as a mirror of human understanding and self-awareness as they have changed over time. Buildings and the landscapes they arose from or contained reflect back to us the image of the people who commented upon them and used them in times past – their sense of themselves.

Underlying the transformation of ideas about nature in Loudon's day were scientific practices aimed at determining the 'true' or essential qualities of living beings, plants and animals. These seem perfectly reasonable to us now, but in the past they were highly speculative, requiring great leaps of the imagination. They required the close observation of living forms and processes with relatively new instruments like microscopes or in fairly novel structures like glasshouses. They included the modelling of chemical, physical or organic phenomena through hypothesis and experimentation. They involved the quantification and calculation of the effects of natural forces and the deduction of their likely causes.[13] These practices underscored the potential Thomas Knight found in the concept of adaptability. In 1805 he wrote:

Experience and observation [author's emphasis] appear to have sufficiently proved, that all plants have a natural tendency to adapt their habits to every climate in which art or accident places them: and thus the Pear Tree, which appears to be a native of the southern parts of Europe, or the adjoining parts of Asia, has completely naturalised itself in Britain, and has acquired, in a great number of instances, the power to ripen its fruit in the early part of an unfavourable summer.[14]

The passage reveals how the order, wholeness or contiguity of the natural world was not simply assumed, but based on the close examination of the distinguishing features of species and, most importantly, their behaviour. The living being – our long-suffering pear tree, for instance – was viewed less as one link in the hierarchical order of creation, as had been assumed for many centuries before Joseph Hooker and Thomas Knight formed their observations. Rather it was conceived of as an organic entity, part of a larger domain joining all living beings and the forces of physical nature – the force of gravity, for instance, the pressure of the air or operations of the atmosphere or the chemical composition of matter. Ultimately, it was upon the body of the organism that the reality of the environment was made apparent. In other words, by observing physiological reactions to levels of light or heat, for instance, or their exposure to certain climates, organisms revealed their dependence on these environmental factors and unique locales. As Knight's comments suggest, they were also capable of being transformed by their environs.

The glasshouse was but one of a number of practical or technical means whereby a new kind of scientific truth – a new kind of visibility – was made possible. It was a catalyst for novel ideas like the adaptation of living beings to their surroundings. It provoked reflection on the consequences of such biological terms for human beings as well as plants. Paradoxically, glasshouses also enclosed a shadowy realm of unanswered questions concerning the limits of life itself and, so too, the limits of human understanding. There, visibility did not always lead to enlightenment, for there was good reason to debate the exact *source* of vitality. Was it simply the force of nature acting on the organism as the sun acts upon the leaves of a plant, or was it something else within living beings that accounted for their ability to adapt to new circumstances or die? Applied to human existence, these questions were immensely important later in the nineteenth century. They elicited environmental concerns among other things. They provoked consideration of human psychology and thoughts on how the mind itself, no less than the body, was influenced by its environs.

In the writings of Michel Foucault, correspondences between ideas and practices, the body and mind, echo and are reified within the nineteenth-century asylum, the hospital and the prison.[15] Georges Canguilhem and François Delaporte studied the consequences of concepts, hypothetical models, scientific and experimental techniques for an understanding of human existence.[16] From a similar perspective Jonathan Crary's work suggests that the history of technology is central to the understanding of the development of human sensibilities and modes of perception.[17] In the historical episodes surveyed by these writers, space is an object of careful scrutiny, a means for articulating a domain of biological realities and a tool for understanding human identity, behaviour and values.

In a similar fashion, the nineteenth-century glasshouses proposed by Mackenzie, the palm houses built by Loudon, Burton and Turner, brought together organic processes and functional architectural forms and contrived landscapes in one confined space in which life and death, the normal and the pathological, were made evident. The glasshouse, to adopt Richard Grove's term and shift his frame of reference slightly, was an 'analogue' of nature. Along with other spaces in which human beings encountered living things in close quarters, the glasshouse enabled the 'newness' of nature to be accommodated, adapted or rendered familiar within certain bounds, all the while stimulating 'an experiencing of the empirical in circumscribed terms'.[18] In making science real, glasshouses, like the Victorian home, revealed hitherto unseen worlds that were novel and provocative. However, as subsequent chapters illustrate, this experience could occur by 'technological default'. By inducing some absence or deficiency – by excluding some vital force or substance – glasshouses, like the Victorian house and garden though in different ways, served as laboratories for studying life. They were also places for cultivating new forms of self-awareness for, in studying the organic world close at hand, one was also led to question the limits of human understanding and power.

To the reader, two caveats accompany the preceding comments. Though citing a number of parallels between developments in biological science and the history of the built environment, this book is not intended to be a conventional biographical or historical account. In making periodic reference to the ideas and projects of John Loudon, its aim is not to imitate or challenge the valuable work of Melanie Simo in her biography *Loudon and the Landscape*.[19] The book's purpose is not to recast Loudon's genius or to re-evaluate his influence. Rather, it follows a particular line of reasoning implicit in many of his writings and evident in others of his time. It is a view that claims that a building or landscape should be designed to function as space for human habitation, just as any organism is housed, as it were, by the environment. The polymath and prolific author of *An Encyclopaedia of Gardening* (1822), like his predecessor, the equally influential and erudite French architect, physicist and natural philosopher Claude Perrault (1613–88), and like others who also found in nature principles to explain human society, give voice to a past less constrained by the boundaries imposed by discrete fields of study than we are today.

Likewise, this book is not another history of nineteenth-century iron-and-glass buildings, domestic architecture or gardens as they are commonly composed: a survey of key works and famous designers or building forms arranged to illustrate social, political or economic formations. The popularity of the glasshouse is presented not as the obvious outcome of some profound desire for the exotic or the need to care for plants, as often assumed. Rather, it is valued for revealing the power of science over nature and the potential for and limits of human control over the vagaries of geography and climate. History, in this instance, as regards the manipulation of living species and inorganic matter, is a means of reflecting on the environment in anthropocentric terms, where perceptions and thoughts on human identity, integrity and character are ultimately the arbiter of value.

In the following pages I return at times to the Palm House at Kew Gardens like someone drawn to the scene of an accident. The reason for doing so is not to ascertain exactly why the palms died, or how they could have been saved if only Decimus Burton, Richard Turner or the Hookers knew what they were doing. Rather, one should ask why those involved with constructing a home for plants were so concerned – why is it, in fact, that hopes for *our* homes are unlikely to be completely satisfied today? Our species may not always come across as well in what follows as in more 'heroic' accounts of nineteenth-century architecture and landscapes, though they should at least prove interesting. It will become obvious that one can draw broad conclusions regarding the beings that peered into glasshouses, built and wrote of them, projected their fantasies and cast their fears upon their 'quaintly twisted pillars' or those who discovered in the home 'innumerable mysteries' portending organic life. When all is said and done, the conclusions drawn from a study of this building type or the home have less to do with them as objects of Victorian fancy, but rather with contemporary desires and anxieties regarding the environs in which we all find ourselves.

The Vital Landscape has two key aims. First, it describes the theoretical developments and historical circumstances whereby human beings came to be seen as both responsive creatures affected by such factors as light, warmth and sound, and perceptive beings whose uniqueness lay in a capacity to derive meaning from their surroundings. It poses the significance of such concepts as organic planning and the functional disposition of space as key instances of the influence of biological or life science on architecture and landscape gardening in nineteenth-century Britain. It notes, among other historical developments, the rise of a specialist structure like the glasshouse in the first half of the century, but then also landscapes like the domestic garden and garden cemetery. It considers these and interior spaces like those composing the house of a British 'gentleman' of the Victorian period in response to newly determined requisites of life and a growing sense of environmental awareness.

Second, this book aims to suggest how landscape features and spaces exemplifying conditions of enclosure and exposure, proximity and distance, interconnectivity and so forth became devices for thinking of ourselves as, in fact, possessing an 'inner' world of motivations, desires and sentiments and as inhabiting an 'outer' world of environmental and social influences. Relations between a visitor to the botanical gardens and its specimens, between a resident in their garden and the outdoors or a retiree and their possessions, were not only the result of so many acts of design, fashioned from old traditions and new fashions. They also served to position the viewers as unique inhabitants of space, beings that derived a sense of self-awareness, uniqueness, wholeness and individuality by reflecting upon the environment around them. The seven chapters that follow are roughly divided in two parts according to these aims. The Conclusion summarizes the main points of the book and offers a few thoughts on related, contemporary concerns.

Interpretations of nature in the past have often been accompanied by a belief in human uniqueness. It has been both presupposed and argued that humankind was somehow different to other living creatures, plants and animals. Because of its rationality it was destined to exercise a mastery over both insensible beings and inanimate matter. As the first three chapters of this book illustrate in various ways, this belief was gradually transformed over the course of the seventeenth and eighteenth centuries due largely to the rise of the life sciences. These chapters identify instances when there seems not to have been anything like an environment and others where something like it begins to appear.

Chapter 1 considers instances of the time-honoured belief in the fullness and ordered arrangement of nature whereby plants, animals and humankind stood in specific relationship to one another. Following an overview of two different views of the natural world in eras prior to our own, it suggests that the equally longstanding opposition between the domains of nature and culture was given renewed attention in late eighteenth- and nineteenth-century European thought. This coincided with the emergence of biological and social sciences that sought to address human identity in new ways. Works of architecture and landscape design came to be valued as evidence of the ways that humankind achieved a wholeness of being by occupying their surroundings in a particular manner. Connecting the way we appreciate the built environment today with that of our predecessors is the useful distinction between these domains of existence. This kind of discrimination has allowed us to find meaning in habitable spaces.

Chapter 2 looks more closely at nature in seventeenth- and eighteenth-century thought, in an era known for the prominence of 'Classical' science whereby plant and animal kingdoms were defined by a mechanical view of creation and human beings by their capacity for self-awareness. This era is contrasted with subsequent developments in biology in the nineteenth century, in terms of new views on organic life and the adaptation of living species. Practices associated with the growth of empirical science span the divide separating these two eras. These include analogous reasoning and experimentation, the forming of hypotheses and deductions and the use of novel measuring devices and instruments. However, the chapter proposes that there was no environment to speak of prior to the rise of biology and attendant disciplines – at least, not in the modern sense of the word.

Chapter 3 revisits the glasshouse, particularly the ideas and historical circumstances surrounding the construction of Loudon's curvilinear glasshouses. It discusses the Palm House at Kew in which ideas about nature were formed and challenged, and anxieties regarding the built environment made evident. Seen within a broader context, the public glasshouses constructed in the middle decades of the nineteenth century in Britain introduced their visitors to other species, animate and inanimate nature in a controlled way, prefiguring in some aspects the closed world of the Victorian house and garden. These were spaces where the novelty of the organic world was introduced and where scientific discoveries could be experienced seemingly within a bounded terrain. As events leading to the demise of the exotics in the Palm House in 1867 suggest, this terrain proved to be much larger than the walls of the glasshouse could contain.

The domain of organic existence in the glasshouse is counterpoised by the Victorian home in the second part of *The Vital Landscape*. Accompanying the rise of empirical science portrayed in previous chapters was a historical process that extended well into the nineteenth century, resulting in more secular attitudes to nature and humankind's place in it.[20] Overshadowing the belief that divine providence had favoured the descendants of Adam with a capacity for choice and freedom of will was a more abstract and scientific account of human identity whereby the species was thought to be endowed with unique qualities of mind. The desire to reconcile an immediate experience of the natural world with a biblical account of human origins was eclipsed by speculation on the nature of the faculties of reason, imagination, perception and memory. Studies of population, natural theology, botany and evolutionary biology alike were complemented by observations made in the domestic sphere of the qualities forming built spaces and landscapes. Both theory and everyday practice drew attention to circumstances of containment and enclosure and their relation to human faculties, physical and mental attributes and forms of behaviour. They encouraged the householder and homemaker to conceive of relations between themselves and their surroundings in both functional and systemic or holistic terms.

Chapter 4 considers how certain desires and anxieties came to be organized through reference to surroundings that were taken to be the source of health and well-being as well as the cause of disease, unhappiness or worry. Through a closer attentiveness to nature's fundamental elements, particularly in the home, one came to be aware of the conditions in which human beings lived, thrived or withered. One consequence of the popularization of science and its claim to certainty in the house and garden was that one could begin to reason about or discern in the environment a myriad of structured or causal relationships between living organisms and earth, air and water. The householder became alert to signs of the influence of organic and physical nature on themselves and those in their care.

Chapter 5 starts with expectations for the systemic order or wholeness of the environment, in both its natural and human-made forms. Works of popular science, treatises of domestic architecture and gardening, household economy and fiction brought investigations into habitats and the geographical distribution of species to a broader audience. The domestic sphere itself became a habitat as important for the well-being of its residents – as meaningful in denoting their 'own place' on the earth – as the tropics, desert or alpine meadow proved to be for the creatures found living in these other environs. The chapter details some of the forms and spatial arrangements that helped homeowners to picture and reorganize their own, immediate place on earth.

Chapter 6 considers the antagonism between 'aspect' and 'prospect' in a popular work of home and garden design, Robert Kerr's *The Gentleman's House* (1864), in which the desire to configure gardens to maximize pleasing views from within the home became problematic given the necessity for suitable architectural enclosure and privacy. Kerr illustrated the need to balance both sets of concerns by imagining a model inhabitant of the house and garden, free to move from one to the other, thereby uniting both into an integrated whole. In this regard, 'landscape' entails a certain positioning of oneself in relation to a range of possible views and spatial experiences, comfort and discomfort alike.

The three previous chapters describe in various ways the 'problem' of life as it was accommodated in the Victorian home. Having been made aware of their responsibility for other species and their dependence on nature's elements in the house and garden, residents were guided by expectations for an orderly existence there, though conflicts inevitably arose. Chapter 7 turns to consider death, the passage of time and memory. The events it details are set against a background of calls in nineteenth-century Britain for the establishment of public parks and botanical gardens to ameliorate the alienating effects of urbanization and industrialization. Calls by reformers for new modes of burial, ostensibly as a means of addressing the morass of decaying corpses growing beneath cities like Glasgow and London, prompted the garden cemetery reform movement in the 1830s and 1840s. Abney Park Cemetery in London is cited as a place where the cultivation of 'solitude' established a particular relationship between the aggrieved, nature and the city. It was likewise a place where park visitors were confronted by a past of noble works and enterprising compatriots and a future demanding equally deserving deeds.

Attitudes to nature in the nineteenth century are notable not for the reason that people began to think *differently* about the environment, but that they began to *think* imaginatively about the environment at all. It was less a consistent application of science or biology and more an ensemble of ideas and a general orientation towards the living world. It was a kind of thoughtfulness that served to make one aware of the context in which human beings live. This was a context formed of the various 'interrelationships between man, his culture and his living and non-living surroundings' past and present. This orientation towards nature and thoughtfulness was a way of appreciating the novelty of surroundings, of questioning their organic and inorganic fabric, and the constraints they imposed on the course of life as well as the opportunities they afforded.

The Vital Landscape raises questions about the reason for or source of human distinctiveness. These questions were not voiced solely in Britain, though the nation provided fertile soil for them to be debated. Drawing on ideas from both home and abroad, Scotland and continental Europe, the influence of its scientists and theorists, from Stephen Hales in the eighteenth century to Francis Galton, Bentham and Loudon in the nineteenth, was to extend beyond the British Isles and well into the future. Circumstances attending the nation's industrial, urban and imperial history provided the British with experimental ground composed of glasshouses and public institutions, houses and gardens to test new ideas and discover even newer ones and to promote them overseas. This ground also brought to the fore lost landscapes and ways of life associated with traditional agrarian or rural practices, real or idealized. These questions and landscapes resonate in our own time of environmental awareness.

Primitive huts and wild gardens

Admiration for the fullness and ordered arrangement of nature has long been a feature of Western thought. Thoughts on 'the great chain of being' guided philosophers in the past to discover the purposeful arrangement of creation, its departments or kingdoms. More importantly, the concept served through various philosophical, scientific or literary means to help individuals position themselves within the overall order of things.[1] Books on architecture and landscape architecture today call up the fullness of nature – more or less metaphorically – suggesting the continuity of human efforts to recreate 'paradise' on earth or to build 'Adam's House' on it.[2] While admiration for nature's plenitude may be longstanding, interpretations of its plan have changed over time – so too has our understanding of human identity, particularly the belief in the distinctiveness of humankind.

The concepts of nature *and* culture have played an important role in supporting this belief. Serving largely speculative or rhetorical purposes, the opposition of these terms has a history as long as a chain of being. It suggests that the human species is composed of either living creatures with basic needs like any other, or social animals with communicative abilities, myths and beliefs that set us apart from other creatures as fundamentally imaginative and creative. Philosophers and other writers have long been fascinated with the figure of the wild man or primitive or the castaway, deprived of human society and forced to apply their skills to subdue nature, as a means of thinking about these two domains of existence.[3]

As Paul Hirst and Penny Woolley have shown, this divide was much heightened in the nineteenth century as new biological and social sciences shed a different light on what it means to be human.[4] Biology posed the question of human uniqueness in a new way. Knowledge, not only of mortality, but of the past – of developments occurring over a lifetime and the evolution of species over the millennia – counterpoised the mere existence of plants and animals as *simply* living things. Self-awareness and the need to know the biological and social contexts of actions both past and present defined our own specific form of adaptive response as thinkers, tool makers and builders. Likewise human individuality and wholeness of character came to depend to a certain extent on just how well people adapted themselves to the environment that science revealed around them.

As ways of knowing the order of things, sciences like biology and sociology encouraged the close scrutiny of the boundaries between their respective domains. Works of architecture and landscape gardening provided evidence whereby the certainties of science as opposed to the vagaries of culture have since been debated. The need for shelter to protect the body from sun, wind and rain or the seemingly timeless physical laws and economies of forces that make structures stand up and endure are commonly thought to be more rationally or objectively known. Reflections on culture, on the other hand, raise issues of aesthetics and expression, ornament and style, beauty and taste. They address values which are more or less relative to each society and impermanent. Of course these oppositions are neither precise nor reconcilable in such terms. This does not mean, however, that they cannot be placed in a historical context.

This chapter identifies two different views of the natural world in the centuries leading up to the rise of modern biology. It illustrates how environmental awareness today was built partly on forms of speculation from the past even though the 'environment' was not itself an object of reflection in earlier times. Buildings and landscapes are cited that exemplify efforts made in the nineteenth century to establish a sound basis for the forms that human habitation *should* take as a means of making design a more rigorous and fundamentally meaningful practice. If one aim unites these efforts it is the desire for certainty, objectivity and truth commonly associated with the exercise of reason.

The purpose of this line of inquiry is not to reconcile philosophically the science of shelter and structure with the aesthetics or 'art' of design. Whether citing the circumstances of 'man' in the jungle or the 'jungle' in man, the 'primitive hut' in nature or the 'nature' of all true architecture, the point made is that each instance presupposes a distinction between humankind and the world it occupies. Each instance leads one to differentiate between environs valued as either accommodating places to live or as places for expressing culture. The environment came into being with just such conjecture, in part. While one may not be able to resolve the dichotomy of nature and culture and the forms of discernment they beg, these entities came to find uneasy accommodation in unique environs, both real and imaginary. Here, appeals to rationality invoke a familiar set of references, valued less for any one idea or precise meaning they may have, but more for their usefulness in debates concerning the distinctiveness of the human species. Reason is a particularly convenient term for opposing nature to culture in defining who *we* are and how *we* inhabit the world.

Forms of irrationality

By defining who 'we' are, the preceding comments should be qualified. Historically, appeals to reason and the nature of one thing or another have characterized Western attitudes and values. To briefly illustrate this point, it is worth considering the impact of early European encounters with people from equatorial Africa, China and the South Pacific in which indigenous buildings were frequently given as evidence of the primitive state of the societies found there. Comments on the primitivism apparent in the architecture of so-called oriental or exotic peoples were the product of both curiosity and profound ethnocentrism. Historical commentaries offer us a glimpse of the qualities Europeans once found desirable in their own habitations, their forms and appearances. These prefigure Western expectations for the built environment today.

In 1669, for instance, the Dutch traveller Jan Nieuhof published an account of his travels to China commonly titled *An Embassy to China* (see Figure 1.1). The work is probably the first Western book to offer an account of Chinese architecture and one of a number to speculate, in effect, on the character of the Chinese mind. Nieuhof observed:

> This Empire is not altogether void of architecture although for neatness and polite curiosity, it is not to be compared with that in Europe; neither are their edifices so costly nor durable, in regard they proportion their houses to the shortness of life, building as they say, for themselves, and not for others … in China they dig no foundations at all, but lay the stones even with the surface of the ground, upon which they build high and heavy towers; and by this means they soon decay, and require daily reparations.[5]

Thirty years later, the French missionary priest Louis Le Comte visited China and found the apartments in the Imperial Palace at Peking 'ill contrived, the ornaments irregular' and wanting uniformity 'in which consists the beauty and conveniency of our palaces'. The seemingly complex forms of Chinese buildings were particularly evident in the 'medley of beams, joists, rafters, and pinions' of the celebrated Porcelain Pagoda at Nanking. In Le Comte's view, this was a defect that proceeded simply 'from the ignorance of their workmen'.[6]

In another instance, in 1793, following the first attempt by Britain to establish diplomatic contact with China, Lord George Macartney furthered these generalized views of its people in tone, if not exact detail. While finding houses that exhibited a degree of magnificence and even some that revealed 'taste and elegance in their decorations', the ambassador discovered at the same time signs of great discomfort and inconvenience. There appeared to be, among other defects, a complete absence of useful furniture.[7] No structure was too insignificant to escape his notice, even the Chinese empire's bridges. Though generally of graceful appearance, they were slight in construction as few allowed wheeled carriages to pass over them. Perhaps, Macartney surmised, this apparent neglect of science as applicable to engineering and other practical purposes was understandable. Given the many millions of Chinese who depended on their labour for survival, their government would naturally discountenance the use of mechanical powers and 'such artificial aid but where it is absolutely unavoidable'.[8] Despite such reasonableness in the art of governance, economy of effort was not found in the limited attention paid to 'Chinese scenography' or the art of laying out pleasure-grounds. Macartney observed that:

1.1 'A Street of Nanking', in Jan Nieuhof, *An Embassy from the East India Company of the United Provinces to the Grand Tartar Cham, Emperor of China* (1669)

There is certainly a great analogy between our gardening and the Chinese; but our excellence seems to be rather in improving nature, theirs' to conquer her and yet produce the same effect. It is indifferent to a Chinese where he makes his garden, whether on a spot favoured or abandoned by the rural deities. If the latter, he invites them or compels them to return. His point is to change everything from what he found it, to explode the old fashion of the creation and introduce novelty in every corner. If there be a waste, he adorns it with trees; if a dry desert, he waters it with a river or floats it with a lake. If there be a smooth flat, he varies it with all possible conversions. He undulates the surface, he raises it in hills, scoops it into valleys and roughens it with rocks. He softens asperities, brings amenity into the wilderness, or animates the tameness of an expanse by accompanying it with the majesty of a forest.⁹

In these passages reason assumes various guises of foresight, familiarity with the function or purpose of buildings and the properties of their materials. It requires principles of rational construction or particular kinds of aesthetic discernment or even prudence and restraint. These were assumed to be the chief attributes of the European mind, the foundation of all positive aspects of European character.

Nieuhof and Macartney reinforced a longstanding view whereby the West offered numerous instances of enlightenment as opposed to other non-Europeans or primitives. They were thought to be motivated solely by base or 'debased' instincts and immediate needs. The British ambassador was able to credit the Chinese of Marco Polo's thirteenth-century *Travels* for having 'reached their highest pitch of civilization'. In his own day, however, and while his own compatriots 'have been every day rising in arts and sciences, they [the Chinese] are actually become a semi-barbarous people in comparison with the present nations of Europe. Hence it is that they retain the vanity, conceit, and pretensions that are usually the concomitants of half-knowledge.'[10]

Evidenced by their buildings and gardens, reason allowed Britons above all others to bring 'order out of the chaos of his natural instincts', or so the nineteenth-century moralist Samuel Smiles believed.[11] The occupants of the British Isles were uniquely suited to relate their building forms to nature because, standing apart from nature, they were able to see it most clearly. They lived alongside nature, knew its laws and worked creatively within the constraints the natural world imposed.

Links in the chain of being

From the time of Marco Polo's first impressions of the Orient to the final decades of the eighteenth century when a Scottish noble felt confident to diagnose the psyche and character, buildings and gardens of the Chinese, two distinct and particularly Western interpretations of the order of things are apparent. These entail two different views of the place of humankind in the natural world. Michel Foucault called these *epistemes*, or conceptual frameworks in which objects of reflection like plants and animals, animate and inanimate objects, were composed.[12] In the Renaissance episteme nature was commonly thought to be a great mystery. It was a domain of being to encompass all other realms of existence.[13] Drawing on Greek philosophy, knowledge of the natural world formed a hierarchy. It was organized to the extent that its branches more or less revealed divine truths. To search for meaning was to uncover the likeness between a particular plant or animal form and the overarching cosmology of which it was a part. Despite uncertainty regarding the fixity of species, humankind was placed firmly above the plant and beast yet below the angels and God. This privileged position was based on human abilities to subdue the former and adore, however imperfectly, the latter.

It was a view whereby the oddities or peculiarities of nature were highly provocative for they had to be made to fit within the divine arrangement of forms. More importantly, they revealed something of the mystery of life. Hence, for the natural philosopher, there appeared fantastical creatures positioned within more than one of nature's kingdoms. Perhaps they drew sustenance from more than one of its departments, the earth, air or water. Consider the Scythian lamb, which grew from a stalk, or the mandrake, a plant notable for having a root resembling the human body. Stranger still, the 'Barnacle tree' was a seashore growth that bore geese at maturity (see Figures 1.2 and 1.3). These creatures often appeared in otherwise prosaic works describing common animals and plants. Concerned with kinds of plant life, books took the form of practical manuals or herbals, chronicles or *histoires* in which plants, gardens or gardening practices were simply described in a

1.2
Frontispiece, in
John Parkinson,
*Paradisi in Sole
Paradisus
Terrestris* (1629).
Parkinson's
book describes
1000 plant
species
cultivated in
England in his
day, their
provenance and
the network of
gardeners in
England and on
the continent
who
corresponded
with one
another and
exchanged
specimens. Note
the 'Scythian
lamb' in the
centre of the
illustration,
growing from a
stalk like a plant

Britannica Conchæ anatiferæ.
The breede of Barnakles.

1.3 'Of the Goose Tree, Barnakle Tree, or the Tree Bearing Geese', in John Gerard, *The Herball or Generall Historie of Plantes* (1633)

narrative fashion.[14] Most evident in these texts in which familiar species coexisted with fantastic ones was a view whereby the body of the living being, its past and present forms, its use and symbolism, were part of a much greater reality. Each living thing was part of

the inextricable and completely unitary fabric of all that was visible of things and of the signs that had been discovered or lodged in them: to write the history of a plant or an animal was as much a matter of describing its elements or organs as of describing the resemblances that could be found in it, the virtues that it was thought to possess, the legends and stories with which it had been involved, its place in heraldry, the medicaments that were concocted from its substances, the foods it provided, what the ancients recorded of it, and what travellers might have said of it. The history of a living being was that being itself, within the whole semantic network that connected it to the world.[15]

For the chemical philosopher or alchemist, while the bonds between non-living things were more or less determined, their forms were not, or so it was believed. The alchemist in particular stood in a special relationship to things. The scientists of today are concerned with the exact observation of physical or chemical phenomena. They form hypotheses and deductions and invent experiments to penetrate the structure of matter. Their medieval and Renaissance predecessors, however, were fascinated by the 'passion', the 'death' and the 'marriage' of substances in so far as they served to transform matter and human life itself. In assuming responsibility for transforming matter or changing the course of life's passions, the alchemist put himself in place of time. Mircea Eliade observed that, for the medieval alchemist,

that which would have required millennia or aeons to 'ripen' in the depths of the earth, the metallurgist and alchemist claim to be able to achieve in a few weeks. The furnace supersedes the telluric matrix: it is there that the embryo-ores complete their growth. The *vas mirabile* of the alchemist, his furnaces, his retorts, play an even more ambitious role. These pieces of apparatus are at the very centre of a return to primordial chaos, of a rehearsal of the cosmogony.[16]

Today, one might regard the divisions between living species and their relationship to one another and to inanimate matter as more or less fixed and obvious. The natural philosopher and alchemist, however, did not necessarily believe this was the case. Albert the Great, born in 1193 and an alchemist as well as Dominican friar who taught Thomas Aquinas, presents an interesting figure in this regard. It is related that Albert once prevailed upon a Dutch nobleman to honour his house with a visit and banquet. The scene was described in 1815 by James Lackington as follows:

Albert had tables laid in the convent garden, although the season was winter, and at that time extremely rigorous; the earth was covered with snow, and the courtiers who accompanied William, murmured at the impudence of Albert, who exposed the prince to the severity of the weather; suddenly the snow disappeared, and they felt not only the softness of spring, but even the parterre was filled with the most odoriferous flowers, the birds as in summer flew about or sung their most delightful notes, and the trees appeared in blossom. Their surprise at this metamorphose [sic] of nature, was considerably heightened, when after the repast, all the softness of the air, the flowers, the singing of birds, the delightful spring ceased – everything disappeared in a moment, and the cold wind began to blow with the same rigour as before.[17]

As they viewed the order of creation both Albert and Aquinas were able to distinguish between its daily operations that were more visible and certain in remaining subject to natural causes than the will of God, which was seemingly unfathomable.[18] Their concerns did not in themselves rule out the observation of physical or chemical matter. Rather, what was different to the science of today was that their investigations relied less on recognizable experimental methods and observational techniques. Their concerns were allied to a different notion of objectivity less likely to divorce theological and scientific interests. It was a notion that demanded particular powers of discernment on the part of the philosopher. Truth was acquired through divination and the discovery of the relations of physical proximity, emulation, analogy or sympathy that connected each being with all others.

In one instance where this view of the world was applied to buildings, particularly the design of centrally planned churches, the revival of ancient Greek and Latin philosophy by Renaissance humanists of the fifteenth and sixteenth centuries cast individuals and their surroundings in a certain light. Seeking to turn the invisible forces, the similitudes or visible resemblances binding all living beings, animate and inanimate matter together to their advantage, architects relied on their knowledge of geometry and proportion to integrate their plans into the fabric of nature. In doing so, they called upon a belief in the involuntary attraction exerted by objects on their viewers and spaces on their occupants. Architects and theorists such as Leon Battista Alberti (1404–72) and Antonio Filarete (1400–69?) built their churches high so that 'those who enter feel themselves elevated and the soul can rise to the contemplation of God'. They covered them with magnificent domes so that the eye 'sweeps round instantaneously without interruption or obstacle'.[19]

This belief in the attractive force of objects and spaces, invoking what is commonly known as empathy theory, could be seen as a precursor to modern thoughts on psychological projection. However, its terms were neither biologically nor sociologically derived. It relied neither on a thorough knowledge of the viewer's cognitive faculties nor their training in aesthetic principles. Incorporating a different view of the connectedness of building forms and the natural world as opposed to, say, the sciences of psychology or behaviourism, humanist architects of the Renaissance nonetheless sought to predict their occupants' experience of space as a designer might today.[20] Their efforts, however, had an entirely different metaphysical basis than modern science. Their churches were not 'environments' in the modern sense of the word, any more than Albert's convent garden could reproduce in exact detail the complexity of the natural world and the myriad of interdependencies that bound all living things to it.

A mechanical universe

Signalled by the criticism of resemblance as a means of ordering things by René Descartes (1569–1650) and Francis Bacon (1561–1626), the influence of ancient Greek thought and, specifically, Neoplatonism waned.[21] During the period characterized as the 'Classical episteme' of the seventeenth and eighteenth centuries, knowledge of living beings was still largely prescribed by the practices that sought to care for them – a context for ideas that persists today. However, attempts were made by natural philosophers to account for the observations which the garden or plantation, hospital or asylum made obvious.

It was an era commonly associated with the rise of empiricism and one in which plants and animals came to be distinguished as objects of study in different ways. Two relatively new fields of inquiry emerged at the time somewhat independent of the practical concerns of horticultural and medical arts. Both entailed comparisons between plants and animals situated within separate, though related, kingdoms. The first, natural history, assigned a certain primacy to plant forms. Perhaps this was so because there were sound economic reasons for cataloguing plant life as the source of food, fuel, clothing and medicine or that plant forms allowed for the extension of powers of description. Plants had the added advantage over animals in that they were easier to see. They were apparently simple and did not walk, fly or slither away from or bite the hand that drew them. With its emphasis on classification, nomenclature and taxonomic technique, the practice of natural history was responsible for imposing sharp divisions among groups of inanimate objects and living beings, particularly between species, though the arbitrariness of these divisions was questioned by some.[22]

Natural history made of this 'heroic' era of science one in which efforts to represent, catalogue and arrange all things have come to betoken human desires to impose a rational order on nature (see Figure 1.4). The influence of natural history was to extend far beyond the study of living species and well into the nineteenth century, forming, as it were, a 'culture' of inquiry that was to influence a broad range of fields, from art and antiquarianism to architecture and gardening.[23] It provided, so to speak, a universal language for ordering form; taxonomical methods occupied minds on both sides of the English Channel and accompanied European colonial exploits abroad as a way of cataloguing potentially useful discoveries. Regardless, in an era of observation, plants came to serve as models for classifying the animal kingdom, thereby assuming a certain 'epistemological priority'.[24]

Easily overlooked by historians of natural history and those interested in sources of nineteenth-century visual culture, the second discipline to emerge during the Classical era was the study of 'vegetality'. It reversed this relationship of plant to animal. The term, of obscure origins according to the *Oxford English Dictionary*, was defined modestly by François Delaporte as the study of 'whatever it is' unique to plants.[25] For the 'vegetalist', vision and understanding were thwarted by the simplicity that guided descriptive or taxonomical studies. In terms of their purposes it was difficult to observe the role played by the leaves,

stems, roots and flowers catalogued by the natural historian (see Figure 1.5). Vegetality entailed the study of physical life and growth and was an antecedent to the science of physiology. It appeared late in the seventeenth century and developed through numerous analogies between plants and animals and the vital processes they had in common. These processes, consequently, formed an empirical realm of existence – they became real.

In 1779 David Hume debated whether or not the world resembled more a plant or an animal than a watch or knitting-loom given its apparent order, spontaneity and regularity. He concluded that the cause of its regularity at least was more akin to that governing plant and animal life. The cause of the world, he inferred, was 'some thing similar or analogous to generation or vegetation'.[26] As 'generation or vegetation' was a force shared by living species, the infinitive verb 'to vegetate' came to denote a domain of physiological operations devoid of consciousness, possibly sensation as well. The conceptual framework upon which the order – if not so much the appearance, in Hume's view – of living beings was established following the collapse of Neoplatonism was 'mechanicism'. It was during the Classical era, according to François Jacob, that

the seventeenth century saw a shift in the center of gravity of the universe. Its world was one in which stars and rocks were subject to the laws of mechanics expressed by means of the calculus. In order to assign a place to living beings and explain their functioning, therefore, there was but one alternative: either living things were machines in which only shapes, sizes, and motions had to be taken into consideration, or else they were not subject to the laws of mechanics, in which case it

1.4 Natural History Museum of Ireland, Dublin, designed by Frederick Clarendon, 1857. The zoological collection, its rooms and exhibits are unique in being preserved largely intact since their establishment in the nineteenth century

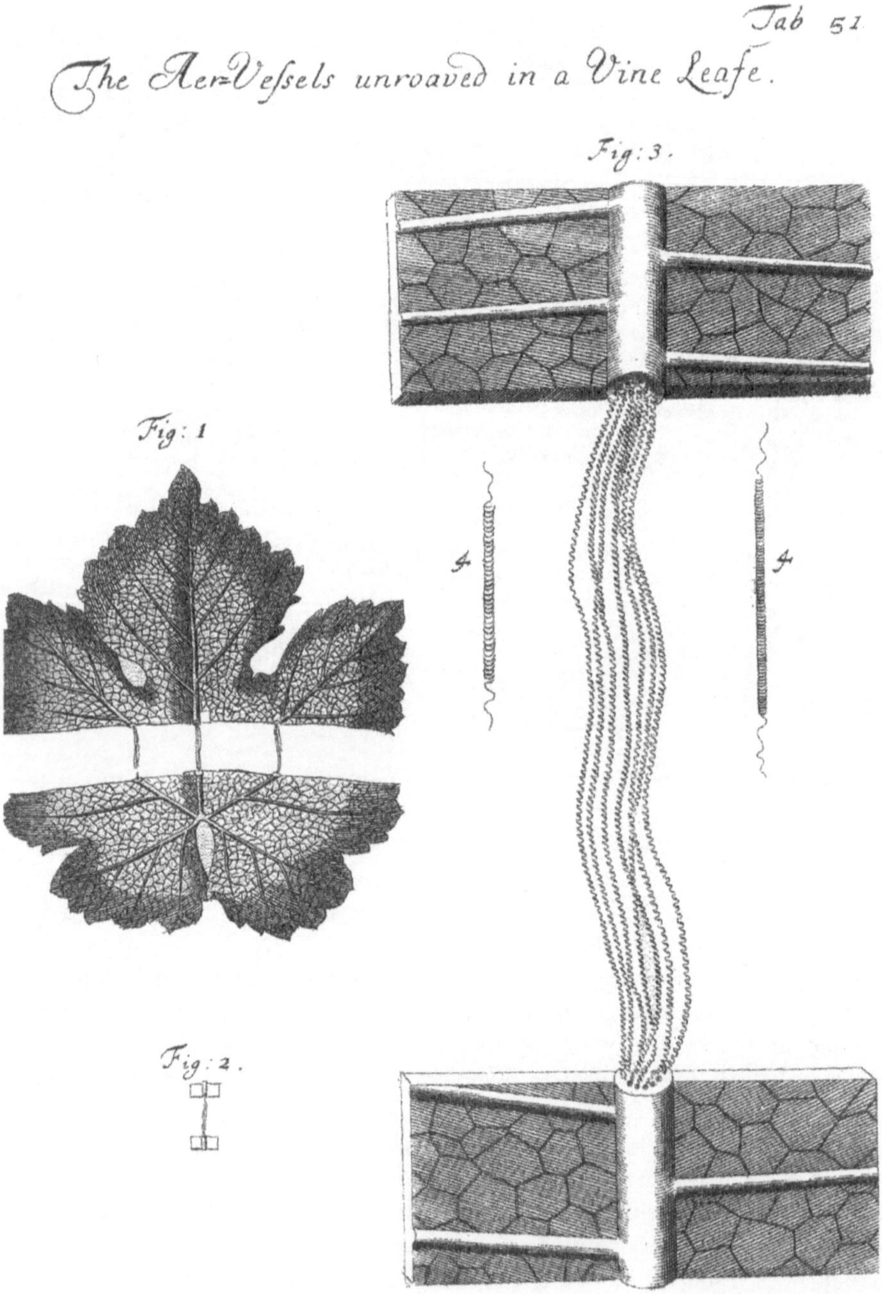

1.5 'Part of a Vine Branch Cut Transversely, and Split Half Way Down the Middle', in Nehemiah Grew, *Anatomy of Plants* (1672). Grew's treatise is notable in being concerned with the intimate structure of plants, rather than simply their outward appearance

Tab 51

The Aer-Vessels unroaved in a Vine Leafe.

Fig: 3.

Fig: 1

Fig: 2.

would be necessary to give up the idea that there was any unity or consistency in the world. Faced with this choice, physicists and even doctors did not hesitate for a moment: all nature was a machine, as machines were nature.[27]

During this period machines like clockworks and looms and other, more experimental devices provided means for thinking about the forces believed to connect living beings to their surroundings. This occurred in a manner substantially different to the efforts both of Renaissance architects and environmental scientists today. By the beginning of the eighteenth century the principle of self-regulation superseded the view of Isaac Newton (1642–1727) that God was the omnipotent unmoved mover of all creation. Gottfried Leibniz (1646–1716) described this animating power as a static property inherent in things rather than as a manifestation of theological concerns. Following Christiaan Huygens's invention of the regulator spiral in 1675, the regulator watch became a common model for imagining the immutability of nature by which a number of disciplines from physiology to economics and politics were posed in abstract terms involving the conservation of matter and the equilibrium of forces.[28] Just as the force of life was no longer thought to come from a divine or supernatural being, neither was it yet thought to be dependent on the spaces that enclosed living species. It was not like, say, the way a plant depends on the closed environment of the terrarium or the animal its native habitat, cases in which both are provided with *all* the necessities required for survival and adaptation. What was clear inside the glass spheres and vacuum pumps popular with seventeenth- and eighteenth-century scientists was not *how* respiration, perspiration or other physiological operations worked together or how an animal might benefit from sharing the same space as a plant. Rather, what was obvious was that nature formed an economy of forces and living creatures were controlled by it.

Given this mechanical view of life one nonetheless finds the persistence of earlier assumptions regarding the distinctive nature of humankind. Carl Linnaeus (1707–78), a figure closely associated with the rise of natural history and taxonomical study and popularized for his empirical studies of nature, was in fact furthering longstanding theological speculation when he penned his aphorism: 'minerals grow, plants grow and live, animals grow, live and feel'.[29] As was commonly held at this time, human beings were distinguished from minerals, plants and animals by the faculty of reason. Humans were superior to animals, or so Etienne Condillac (1714–80) believed, for they possessed a mental capacity that arose solely from an ability to receive, store and utilize sensations formed of the natural world. Following John Locke (1632–1704), Condillac believed that knowledge of nature began with first, simple ideas which formed the mental images of sense impressions. These then combined to form more complex ideas such as the concepts of order and beauty. Humans were even further distinguished from animals in that they possessed souls. Along with reason came responsibilities stemming from self-awareness and, particularly, a moral imperative to care for the less fortunate members of creation. The mathematician and philosopher Jean-Antoine Condorcet (1743–94) believed that botanists, more so than other scientists, were disposed to piety, for 'Observation of the plant kingdom seems to call up more forcefully the idea of a first cause, to tell us more about its boons and to incline our soul more naturally to gratitude.'[30] Care for nature's creatures, as Delaporte had observed, was intimately connected to their exploitation. For those scientists

opposed to the notion that plants are in some way sensitive emphasized the fundamental goodness of nature, or again, of God, who would not allow defenceless creatures to be made to suffer. Those who believed in the existence of sensitivity in plants invoked a God wholly preoccupied with the balance of nature, for whom suffering had its *raison d'être*. In actuality, the point in both cases was to exonerate man of guilt as a destroyer of plant life.[31]

The discourse of vegetality and the language of machines deserve to be considered more closely. The view of the world they entailed provides a counterpoint to biological science and the idea of the environment. Though each era should be considered in its own historical context, spanning the conceptual divide that separates them was a belief in the distinctiveness of humankind. Common to both periods was the use of technical means to position individuals, plant, animal or human species within their surroundings.

Wild living

The philosophical and theological values that distinguished humankind from other living beings for centuries were subject to close scrutiny with the rise of empirical science. There followed efforts to determine the limits not only of natural forces upon the mind and body, but the impact of cultural factors as well.[32] In the eighteenth century, writers made experiments out of a number of curious incidents involving the discovery of wild or feral children abandoned in the wilderness and deprived of human contact as a way of questioning the bounds of nature and culture. Authors composed works which overturned the privileged place of the human species in the scheme of things in order to consider their moral fibre and behaviour more closely. For instance, in the last of four books comprising the popularly titled *Gulliver's Travels* (1726) Jonathan Swift introduced his readers to the character of the wild, sub-human Yahoo and an archetypically civilized race of horses, the Houyhnhnm, a species 'so well united, naturally disposed to every Virtue, wholly governed by Reason'.[33] Other writers relied on the novelty of unfamiliar settings, depicting Europeans shipwrecked on remote islands and forced to apply superior skills to the taming of wild nature. Daniel Defoe's protagonist in *Robinson Crusoe* (1719) is one case in point. Crusoe's relationship with Man Friday allowed readers to speculate on the commonality of human experience, where basic needs like shelter were more or less fixed, or to consider how experience was mitigated by social circumstances. More recent popular fiction and contemporary film abound with similar stories, from Johann Wyss's *Swiss Family Robinson* (1812) to interpretations of Pierre Boulle's story *Planet of the Apes* (1964).

For eighteenth-century writers the study of language was a useful measure of humankind's privileged status. This was so for Jean-Jacques Rousseau (1712–78), commonly regarded as the originator of the concept of the 'noble savage' and intent on elaborating the way in which human society progressed from a state of primitivism.[34] But, as Hirst and Woolley have pointed out, for Rousseau the lone figure in a state of nature was

anything but noble. Rousseau argued that man's current attributes are the product of a long history of association and that postulating the 'state of nature' is a kind of exercise in abstraction, a stripping away of all that man owes to social relations. Natural man is a creature of instinct and appetite, without language or self-consciousness. He is solitary and wild, living in an eternal present, without ideas and subject to his immediate needs.[35]

Others have used similar terms to describe Rousseau's portrayal of nature as an idealization. It was the 'portrayal of spontaneous equilibrium between the world and the values of desire, a state of prehistoric haphazardness' through which the specific acculturated being of humankind is reflected upon, and not an actual state of existence.[36] In other words, the state of the primitive, like the opposition of nature and culture, is neither innate nor immutable. It is derived not from one essence or source of spirituality, but rather serves speculative or rhetorical purposes. It was for speculative purposes that in 1726 Swift, along with many of his contemporaries, took keen interest in the arrival of Peter, a 'wild boy' from the forests of Germany, at the court of King James. Defoe later wrote of him (1787):

He is now ... in a State of Meer [sic] Nature, and that, indeed, in the literal Sense of it. Let us delineate his condition, if we can: He seems to be the very Creature which the learned World have, for many Years past, pretended to wish for, viz. one that being kept entirely from human Society, so as never to have heard any one speak, must therefore either not speak at all, or, if he did form any Speech to himself, then they should know what Language Nature would first form for Mankind.[37]

Through language the civilized human being was thought to have been made aware of what distinguished it from lesser creatures and was able to articulate the difference between itself and other species. This distancing effect of language, furthermore, was coupled to another. It was a kind of power derived from innate skills and technical abilities whereby humans were able to escape their primitive surroundings by transforming nature into something else – some *place* else – entirely. Seen in this light, Edgar Rice Burroughs's fictional character Tarzan (1912) was indeed king of the jungle, for he alone was capable of tearing the jungle down.[38]

Primitive huts

Shifting attention from thoughts on wild men to the houses they lived in, Marc-

Antoine Laugier (1711–69) is the figure most often considered to have brought Rousseau's critique of primitivism to bear on the study of the origins of building. In his 1753 work, *Essay on Architecture*, he reflected upon the origins of architecture in the form of the first primitive hut (see Figure 1.6). In the frequently quoted introductory chapter to his *Essay*, Laugier invited the reader to consider

man in his primitive state without any aid or guidance other than his natural instincts. He is in need of a place to rest. On the banks of a quietly flowing brook he notices a stretch of grass; its fresh greenness is pleasing to his eyes, its tender down invites him; he is drawn there and, stretched out at leisure on this sparkling carpet, he thinks of nothing else but enjoying the gift of nature; he lacks nothing, he does not wish for anything. But soon the scorching heat of the sun forces him to look for shelter. A nearby forest draws him to its cooling shade; he runs to find a refuge in its depth, and there he is content. But suddenly mists are rising, swirling round and growing denser, until thick clouds cover the skies; soon, torrential rain pours down on this delightful forest. The savage, in his leafy shelter, does not know how to protect himself from the uncomfortable damp that penetrates everywhere; he creeps into a nearby cave and, finding it dry, he praises himself for his discovery. But soon the darkness and foul air surrounding him make his stay unbearable again. He leaves and is resolved to make good by his ingenuity the careless neglect of nature. He wants to make himself a dwelling that protects but does not bury him. Some fallen branches in the forest are the right material for his purposes; he chooses four of the strongest, raises them upright and arranges them in a square; across their top he lays four other branches; on these he hoists from two sides yet another row of branches which, inclining towards each other, meet at their highest point. He then covers this kind of roof with leaves so closely packed that neither sun nor rain can penetrate. Thus, man is housed. Admittedly, the cold and heat will make him feel uncomfortable in this house which is open on all sides but soon he will fill in the space between two posts and feel secure. Such is the course of simple nature; by imitating the natural process, art was born. All the splendours of architecture ever conceived have been modelled on the little rustic hut I have described.[39]

Laugier's treatise set out to renew the basis for building by identifying fundamental principles. These were based on an experience of nature whereby the need for shelter was mediated by the self-consciousness of the primitive. Writers have commented on the popularity of Laugier's theory of origins, particularly in the second half of the eighteenth century among British architects like Sir John Soane (1753–1837).[40] Though Laugier's theory was called into question early in the nineteenth century, a view of the organic relation of human beings and their houses to the natural world illustrated in his *Essay* grew increasingly commonplace. Its significance as an *exposé* on human identity, though, is worth elaborating.

What comes across in the preceding passage, though admittedly in rough outline, is a notion of human adaptability. It is a concept that will achieve its fullest expression in the work of Charles Darwin and his disciples in the following century. Laugier's primitive builder is one driven by an essential need for comfort, while the savage's relationship to his surroundings is cast as largely antagonistic. Of the primitive, Rousseau argued that his 'strength is strictly proportionate to his natural needs and to his primitive state'.[41] Something is always missing from his surroundings, as architecture emerges out of a growing awareness of the meaning of absence and the sensations it

1.6
Frontispiece, in
Marc-Antoine
Laugier, *Essai
sur l'Architecture*
(1755)

produces. A sense of the sun's heat led the primitive to seek shade, of humidity, dryness; a sense of suffocation compelled him to find a means of ventilation and so on. As the primitive's relationship to nature is construed as organic, this requires mediation through built form. In Laugier's narrative, in references that reveal the author's own European origins as well as his anthropocentrism, human nature would seem to have been driven by the need for square plans and architectonic forms. Paradoxically, while 'stripping away all that man owes to social relations' in the form of building, ancient architectural orders, Classical proportions and ornament, these two traces of Western culture remain. The dwelling encloses a domain in which building acquires meaning as fulfilment of basic needs. Substitute Laugier's 'little rustic hut' with its counter-image in Le Comte's condemnation of the Porcelain Pagoda or Macartney's indictment of Chinese pleasure-grounds, and one finds a similar story told. It is a tale of the unique suitability of the European to build *with* nature, all the while standing apart from it.

The influential theorist and academician Jean-Nicolas-Louis Durand (1760–1834) took exception to specific aspects of Laugier's theory though the practice of speculating on the origin of building was shared by several continental writers at the beginning of the nineteenth century, appearing in the writings of the modernist architect Le Corbusier almost two centuries later (1931).[42] A contemporary of Durand, Antoine Quatremère de Quincy (1755–1849) came to reject the monogenetic basis of Laugier's *Essay*. This was the belief that all true architecture evolved from only one source. De Quincy developed his own theory of epigenesis, postulating that architecture grew from multiple foundations. These were the huts of the Greeks, the tomb-like structures of the ancient Egyptians and the tent forms used by the Chinese.[43] This view had important implications for contemporary architectural theory. It not only introduced a degree of relativism into the study of the history of building, but also furthered an awareness of the impact of environmental factors – at the very least, geographical circumstances – on building forms. This sense is evident in *A Discourse on the Origin of Inequality* (1755), where Rousseau observes:

In proportion as the human race grew more numerous, men's cares increased. The difference of soils, climates, and seasons, must have introduced some differences in their manner of living. Barren years, long and sharp winters, scorching summers which parched the fruits of the earth, must have demanded a new industry. On the seashore and the banks of rivers, they invented the hook and line, and became fishermen and eaters of fish. In the forest they made bows and arrows, and became huntsmen and warriors. In cold countries they clothed themselves with the skins of the beasts they had slain.[44]

Speculating on primitive huts and the cares they assuaged gave few clues as to what the built environment *should* look like. However, differences in manners of living would prove increasingly important in establishing a link between a sense of habitat that was to emerge in the nineteenth century and a belief in national styles of architecture and landscape gardens. Serving speculative purposes, primitive huts were useful given that they were abstract and devoid of detail. In other words, they were blank slates for projecting Western values, needs and cultural forms. Primitive huts were no more environments than were Renaissance churches and the glass spheres of the vegetable scientist. They were influential, nonetheless, in drawing attention to the meaningfulness of the earth's habitation by humankind.

Excesses of art

Following the preceding references to industrious savages and primitive huts, a parallel can be drawn between the longstanding opposition of nature and culture, on the one hand, and the common practice of opposing utilitarian and expressive aspects of designs, on the other. Of course the latter practice was not initiated by Laugier and his followers. Even for Vitruvius the purpose of a building was one quality to be clearly distinguished from a number of others. Likewise, a survey of attitudes to the built environment would easily show that this distinction is neither itself static nor is it the only basis for 'being in the world'.[45] These projects aside, it seems worth considering how the opposition of nature and culture, utility and expression, has proven extremely useful in trying to determine what can be known about works of architecture and landscape in terms of how they might be made and lived in.

Importantly, such distinctions were heightened in the nineteenth century by writers who found in buildings and gardens evidence for new ideas about the fullness and interconnectedness of nature, where laws governing animate and inanimate matter were distinct from those arising from social relations and human values. Authors of domestic treatises reinforced a broad ethical imperative to relate the integrity and character of forms and spaces to their inhabitants, desirous to possess similar qualities themselves. This imperative involved a particular regard for the purpose of spaces. It called into play the natural historian's references to formal character, type and classes, shifting their frame of reference from living beings to buildings and landscapes. It demanded the imaginative use of new materials and construction techniques to create spaces for an expanded array of purposes. These circumstances characterize the nineteenth century in particular as a period in which the act of building became a very complicated process indeed, involving an ever-increasing number of fields of expertise and manners of discernment.

Novel buildings and true architecture

Historical accounts of nineteenth-century architecture provide evidence of this ethical imperative. Novel structures such as iron bridges and railway sheds, exhibition halls and glasshouses have provided historians with means by which building practices accompanying the rise of science and industry have been distinguished from practices of ornamentation derived from social instincts. Such structures have served as devices, akin to literary tropes or figures of speech, as it were, for asserting the identity of humankind as builder. In many accounts we see ourselves reflected back as beings driven to transform the natural world in order to inhabit it. Nikolaus Pevsner, for instance, wrote of the constructional necessity of certain nineteenth-century forms in terms of nature and culture. He located a particular source of modern design not in any building at all, but upon a number of fixed-arch and suspension bridges. Referring to Isambard Kingdom Brunel's Clifton Suspension Bridge at Bristol (1836), he wrote:

It seems hardly admissible that the beauty of such a structure should be purely accidental, that it is the outcome of nothing but intelligent engineering. Surely, a man like Brunel must have been susceptible to the unprecedented aesthetic qualities of his design – an architecture without weight, the age-old contrast of passive resistance and active will neutralised, pure functional energy swinging out in a glorious curve to conquer the 700 feet between the two banks of the deep valley. Not one word too much is said, not one compromising form introduced. Even the pylons are entirely unadorned, admittedly against Brunel's original Neo-Egyptian decoration. As they happen to be, they form a superb counterpoise to the transparency of the iron construction. Only once before had such *daring spirit* ruled European architecture, at the time when Amiens, Beauvais, and Cologne were built.[46]

As the story unfolds in this passage, the structural integrity of the bridge makes it a necessary and obvious link with the laws of physical nature. This is counterpoised by its ornamented pylons, representative of a culture of stylistic eclecticism. The so-called 'daring spirit' of Brunel's day provides a link between the past – a European past – and the historian's valuation of reason, innovation and self-restraint as the basis of true architecture.[47]

A contemporary of Brunel, the architect and academician Gottfried Semper (1803–79) was to embrace the challenge posed by industrialization and new building techniques in his day and his influence was to extend far beyond the world of his German readers. Semper's work has been cited as a source of ideas regarding functional design. Like Pevsner, he too counterpoised seemingly timeless principles of design and construction to the vagaries of human society. Semper was living in London at the time of the construction of Paxton's Crystal Palace and the building's impress upon his *Four Elements of Architecture*, which he was writing at the time, is clear.[48] In the Preface to that work, Semper recorded his admiration for nature with its seemingly infinite variety of forms, all derived from a very few 'normal' ones, types or formal characters. He argued that true architecture was based not simply on the imitation of natural models, but was essentially original. It was based on an organic relationship between the rule of law and internal necessity likewise found in nature.[49] Four elements or 'motives' accounted for the continuous evolution of built form. These were, in effect, basic ideas about habitation linking the 'high' art of his day with the first primitive hut. The first, the hearth, was the nucleus of social interaction. This was protected from a potentially threatening natural world by the remaining three elements: the mound, the roof and the vertical enclosure. Continuity of purpose and variability of expression were further provided for in Semper's representation of fundamental construction techniques and details such as weaving and the seam or knot. Having appeared to serve practical purposes, these details soon acquired symbolic value for humankind. They moved, in other words, from the domain of necessity to one of meaning. Thus, Semper was able to celebrate both the novelty of the Crystal Palace and its rightful place in the evolution of built form. Glass now covered iron in the modern exhibition hall, just as straw matting was once drawn across the wooden frame of the primitive hut. While providing serviceable accommodation, the need for wide, column-free spans, light and translucent surfaces, the hall could also accommodate its occupants to their own past as well as accustom their architects to the value of innovation and progress.

Just two years before the completion of the Crystal Palace, James Fergusson entered the rising debate on the 'art' of architecture with a monumental work of his own, published in numerous editions and bearing all-embracing titles such as *Handbook of Architecture* (1855) and *History of Architecture* (from 1865 onwards). In an introductory essay published in the first 1849 edition Fergusson included a diagram counterpoising an 'ordinary house, such as is found in many of our London streets', and a work of true architecture (see Figure 1.7).[50] To the left was drawn the most prosaic form of building, being little more than a wall with openings for windows. Moving to the right, a progression of elevations illustrated principles whereby building form and details gradually evolved. These seemingly arranged themselves to 'produce a more agreeable effect'. Cornices and brick courses expressed the differentiation of floors. In the third section of the diagram the rustication of stonework on the lower storey afforded a greater impression of stability. Meanwhile, the further accenting of windows and roofline with mouldings and pediments and the addition of a parapet carried the house 'out of the domain of building into that of architecture'. The fourth stage in the evolution of the design extended the architect's concern to relate one element to another by further accentuating its ground and attic floors and by imposing a similar division into bays across the façade. In Fergusson's diagram buildings to the left were formed by basic requirements for shelter and the rational arrangement of form and structure. To the right architectural forms emerged as these arrangements were mediated by expressions of cultural concerns, ornament and style. Through such means Fergusson cast himself as a uniquely skilled designer who knew not only how to build, but also how to create forms that were both purposeful and inherently meaningful.

Wild gardens

The history of gardens provides yet further instances of the usefulness of opposing nature to culture. The 'naturalness' associated with a uniquely English style of gardening in the eighteenth century, for instance, was promoted by the poets Addison and Pope. They championed the mind's independence in being allowed to explore a variety of freely arranged landscape features so that associations might be formed between objects and

x -----A-----x-----B-----x-----C-----x------------D------------x

No. 2.

1.7 Diagram illustrating the 'art' of architecture, in James Fergusson, *A History of Architecture in All Countries from the Earliest Times to the Present Day* (1893, ₁1849)

ideas. As John Dixon Hunt has argued, this belief in the freedom of association helped establish a link between the English garden and a more general sense of liberty.[51] It was accompanied by condemnation of the apparent artifice of topiary and the planting of geometrical garden beds associated with continental horticulture and, consequently, the call for the Englishman's liberation from undue foreign influence. John Loudon derided the French for their 'excess of art', for the confusion wrought by gardens filled with 'too many walks; by too many seats and buildings; and by too meagre a distribution of trees and evergreen shrubs'.[52] Behind his criticisms of the malformed state of French gardens that brought confusion to the eye was a veiled condemnation of Gallic character, of the French for being slaves to fashion.

Appreciation for the arrangement of plant species in their natural state and the desire to limit human artifice guided the work of William Robinson, author of the best-selling manual *The English Flower Garden* (1883). Robinson developed an alternative to highly contrived, extensively bedded and costly managed grounds: the 'wild garden' (see Figure 1.8). Robinson took pains to distinguish his work from the picturesque, which proved costly to implement and which generally ignored the opportunities afforded by the discovery of botanical beauty from far afield. His followers sought to place 'plants of other countries, as hardy as our hardiest wild flowers, in places where they will flourish without further care or cost'.[53] The wild garden provided a justification for appreciating the variety, if not the inviolability, of nature. It was not an early work of ecology per se, but one in which species variation served the improving and time-pressed hand of the British gardener. The design of the wild garden acknowledged the geographical specificity of botanical species, their hardiness derived from having grown in unique regions, situations and climatic zones as well as the pleasure to be had in

1.8 'Colonies of Poet's Narcissus and Broad-Leaved Saxifrage', in William Robinson, *The Wild Garden* (1894, [1]1870)

their careful placement. His 'bold' attack on what was considered 'unnatural art' in garden design was contemporary with the appearance of handbooks that popularized British native flora, flowers and plant names. These established an etymological link between the 'naturalness' of indigenous species and their connection with British geography, culture and history.[54]

Conditions of life

The values accruing to the idea of 'wild' or untouched nature versus the obviously artificial or contrived landscape guided Loudon's efforts to distinguish his own gardening style from those commonly labelled picturesque. Variously termed the 'Natural' or 'Characteristic' and then in 1832 the 'Gardenesque' style, Loudon developed thoughts on the integrity and systemic arrangement of plant specimens which botanists were just beginning to explore. He advocated, amongst other practices, that a plant, tree or shrub should be allowed to grow freely so that its spontaneous development could be appreciated by the spectator. He proposed that objects in the landscape should not be considered abstractly as merely sculptural forms, but in relation to others near them and that houses should not be separated from the landscape, but rather harmonized with it.[55] There was obviously much latitude in how to implement these principles, there being many different ways to group plant species or to achieve something like 'harmony' between residence and grounds. Hence, Loudon's advice was, by and large, ambiguous. His experiments with curvilinear forcing-houses raised similar questions concerning the integrity of buildings based on their provision of shelter versus their overall appearance. Drawing the reader's attention to the key difference between scientific and aesthetic concerns, Loudon wrote in 1817:

Imagine, instead of a row of glazed sheds, a row of detached sections of spherical bodies of an almost perfect transparence – the genial climate and highly coloured productions within, obtaining during the whole day the unobstructed influence of the sun's rays, and the construction of the edifice combining the greatest strength and durability – what will be the expression?[56]

By invoking the domains of nature and culture, by speculating on how best to represent the former and avoid an 'excess' of the latter, works and designs by Loudon, like those of many of his contemporaries, brought into play a view of the built environment as a context for discernment. Buildings and landscapes were a medium for expressing the designer's own integrity in honouring nature's 'spontaneous' arrangements whereby each species found its own home. They were likewise means of expressing the designer's individuality in being able to perceive those arrangements and give them form.

When discussing the planning of gardens, the disposition of villas and workers' housing and particularly the design of certain 'expert' or novel buildings like glasshouses, Loudon sought to establish a link between the diversity of building forms and practices in the past and in his own day. He succeeded in this by outlining fundamental principles regarding their inherent *usefulness* as habitable places. He asked his readers to distinguish historically determined styles such as the Gothic or the Classical – even the Chinese or the 'Hindoo' – from building forms and landscapes designed for the sake of necessity. He counterpoised the legitimacy that society bestows upon ancient authorities like Vitruvius and the greater relevance of principles derived from observing the course of human history. He simply presupposed a distinction between the apparent objectivity of the need for habitable space and the relativity of ornament and style; hence his condemnation of John Nash and the Royal Pavilion at Brighton for its 'mimicry of nature'. Distinguishing between the need for shelter and aesthetic concerns, Loudon went further. He attempted to account for the vagaries of social practices philosophically, by calling upon then current theories of 'associationism'. These were evinced when he wrote: 'A barn disguised as a church would afford satisfaction to none but those who considered it as a trick. The beauty of truth is so essential to every other kind of beauty, that it can neither be dispensed with in art nor in morals.'[57]

George Hersey attributed this philosophical tradition to the 'common-sense' school of David Hume (1711–76) and Francis Hutcheson (1694–1746) as developed by Dugald Steward and Thomas Reid and appearing in Archibald Alison's *Essays on the Nature and Principles of Taste*, first published in 1790.[58] This theory held that truth could only be known through the common experience of the ordinary man. Beauty was not intrinsic in objects but existed only in the mind and perception of the beholder. It followed that the purpose of education was to equip the minds of individuals with as many of these associations as possible. These encouraged the student, designer, or gardener in this instance, to cultivate a higher level of aesthetic appreciation.

What was cultivated in the minds of Loudon's readers? Likewise, what powers of discernment were encouraged by Semper, Fergusson and Robinson – even Pevsner, for that matter? Following the advice of these commentators, it was an awareness that some good was to be had, namely the integrity associated with providing suitable shelter or structure or even 'agreeable' effects, when *designing* with nature. It was the understanding that, by providing for 'whole' and wholesome accommodation based on the seemingly universal need for habitation and nature's laws, one might find space to enjoy or fight over what was left in a building for the expression of culture. In Robinson's *The Wild Garden* it was the prospect of working with nature to find both variety and ease in the arrangement of botanical species.[59] Pevsner provided the clue that nature might in fact be the inspiration for a new way of building, functionally and with restraint. This is perhaps the motivation that led his contemporary, Reyner Banham, to distinguish between the use of technology in 'conscious architecture' and in vernacular building.[60] Each of these writers in their own way drew their reader's attention to the importance of the built environment as a framework for decision-making. They introduced readers to ways whereby they might live alongside nature and derive meaning from it, though none were 'environmentalists' in the modern sense of the word.

It is commonly asserted that reason is about knowing and dominating nature. If this is true, then the built environment has proven to be one means whereby we have turned the apparent order of nature to human purposes. Types and classes of buildings and gardens, whether imaginary or real – in the form of primitive huts, glasshouses or crystal palaces, Chinese pleasure-grounds or wild gardens – have served, in fact, to define those purposes. They have allowed us, however imperfectly, in theory and in practice, to fulfil our needs for shelter as living creatures like any other. They have allowed us to assert our own uniqueness, to dress up our houses with style or moderate excess in the garden. They have allowed both house and garden to become expressive and meaningful by design. Invoking ways of knowing things and so too ourselves, Hirst and Woolley use the term 'policing' to describe the negotiation of boundaries between biological facts and social necessities. While, logically, one may not be able to reconcile nature with culture, they are accommodated, if not reconciled, by the idea of the environment and thoughts on the imprecise line dividing natural and human-made domains of experience. Less important for signifying an object or region or a state of enclosure, the environment was more a context for thinking about – for establishing and negotiating divisions within – what George Romanes, a follower of Charles Darwin and author of 'Animal Intelligence', described in 1881 as 'the sum total of the external conditions of life'.[61]

Vegetables in forcing-houses, humans in glasshouses

In the seventeenth and eighteenth centuries the study of nature was guided only partly by abstract speculation. Both the natural historian's invention of a universal language of living forms and the vegetalist's use of analogical reason to understand how living things 'worked' could only partly explain the natural world. Nature was also studied using novel devices and experimental instruments intended to isolate its different aspects, its constituent elements or forces, in order to explain plant and animal physiology or to articulate the perceptive abilities of humankind. The use of such tools was an important catalyst for novel attitudes towards the natural world. Though varying in form, they remain a medium for articulating environmental concerns today.

Some of the devices used by scientists are particularly interesting, for they isolated species within discrete enclosures as a way of examining their vital processes. Some even mediated between human beings and their animate and inanimate surroundings. These were not environments as such, for reasons soon to be made clear. Rather, they served to link aspects of seventeenth- and eighteenth-century science to modern biology, via a set of terms and techniques for enclosing and manipulating space. Along with other 'real' spaces in ships, mines and factories, or the cottages of agricultural workers, the scientist's experimental confines provoked broader questions concerning the need for space on the part of the residents of the British Isles and its impact upon them. They prefigure the containment of nature in the glasshouse and, more generally, in a number of buildings and landscapes.

This chapter begins by revisiting the archaic science of vegetality, the study of qualities unique to plant life. It questions its lingering influence in the early nineteenth century when Classical mechanics and the language of machines were used to place plants in a remodelled version of the forcing-house. Traditionally forcing-houses were simple horticultural structures. They were commonly formed from window sashes placed on low timber boxes or against masonry walls and angled to permit the sun's light and heat to reach plants inside (see Plate I). Assertions of the forms that they *should* take and by implication the space they enclosed as well, of the kind offered by George Mackenzie in 1815, were associated with experiments to hasten the germination of plants and lengthen growing seasons. Apart from these specific purposes, the redesign of the forcing-house followed a century of investigations where plants and animals were placed in enclosures of various sorts to question how their respiration, perspiration or nutrition depended on the earth, air and water around them. In a way, by countering the seemingly automatic workings of plant and animal physiology with the human being's capacity for self-awareness, Classical science reinforced a longstanding view of human distinctiveness through experimental or technical means. In other words, in being aware of their surroundings, humans were not contained or explained by the glass tubes, spheres and forcing-houses that were built to enclose lesser beings.

This point introduces a second and somewhat different kind of enclosure to be explored in this chapter. As a model of perception, if not self-consciousness, the camera obscura was the human equivalent of the forcing-house. It was a tool, like its counterpart the camera lucida, for representing a particular view or composition of objects on a flat surface, allowing them to be drawn in a certain way.[1] Long popular with artists, the camera obscura formed a small chamber that excluded all sources of illumination except for rays of natural light that entered through a small aperture in its side. The aperture allowed images of the outside world to be cast upon an interior surface of the chamber. Though some formed rooms big enough to hold several observers, others had hardly room for one person. None was meant to be occupied for any considerable length of time (see Figure 2.1).

Apart from its usefulness for artists, the camera obscura served more abstract purposes, furthering speculations on perspective and the representation of reality, optics and vision. Even today these now antiquated devices remain popular with theorists seeking to relate perception to historical and cultural circumstances or to discern forms of visuality in the history of photography, film and art.[2] In his often-cited *Essay Concerning Human Understanding* (1690) John Locke used the image of the camera obscura to diagram the human sensory apparatus and to describe the role of vision in connecting a sense of one's surroundings to understanding. In a way it was portrayed as an extension of the human's visual apparatus. As such, it was a diagram for thinking of human beings as possessing an inner world of sensations and as occupying an outer world of things waiting to be experienced.

Both the scientifically designed forcing-house and the philosopher's camera obscura were conceived in functional terms. They recorded or otherwise exploited the forces coursing between their living occupants and the natural world. They made obvious the actions arising from the latter and the reactions exhibited by the former. While they enclosed 'spaces', there was no space for thinking about the environment in either Locke's darkened room or the forcing-house. The species that occupied them were influenced by their surroundings, but in a very limited way. It was a constrained view of the natural world imposed by the philosopher's model of sensibility and understanding. Nature's influence was narrowed into fine beams of light that worked on the human mind as though it were a cog in a machine. Equally,

2.1 Eighteenth-century engraving, 'Use of the Camera Obscura'

there was little room in the camera obscura for the idea of the inhabitant either. In practice, however, by revealing what was missing from the closed world of the laboratory or Locke's imaginary apparatus, these devices provided a vantage point for considering a much larger and far more complex world outside.

One should reflect on the machinations of seventeenth- and eighteenth-century science, vegetality and mechanicism. They not only reveal a collection of seemingly bizarre ideas about how the world was once assumed to work, but also hint at the origins of a more contemporary appreciation of the environment's fullness and constituent parts and, perhaps most importantly, its complexity. The faith that a vegetalist like Stephen Hales placed in machines as descriptive of nature's arrangements contrasts with a sense today that we can never fully understand or adequately re-arrange the environment to suit human purposes. In either case the tools of empirical research that once served to position scientists like Hales and his contemporaries as their subjects in relation to the world around them continue to do so now. While we may not live in forcing-houses, much less in small darkened rooms, other spaces remain for us to situate and reflect upon ourselves and our co-habitants upon the earth.

To 'number, weigh and measure'

During the eighteenth century the study of one of nature's kingdoms depended on the others. When examining plants, scientists had at their disposal knowledge of animal functions. In studying animals, humankind found a substitute for questioning its own unique identity.[3] In this and the preceding centuries, plants and animals came to be distinguished as objects of study in new ways. Experimental devices, instruments and machines were important for they were both models and measures of 'how' living things worked. If arms and legs were thought to be analogous to a mechanical lever, for instance, it was because the ensemble of bone, muscle and tendon looked like a lever. Together they worked to achieve the same result. Similarly, if stems and leaves appeared to move in a manner akin to the limbs of the human body, then it was believed that there must have been some component of the plant which acted like muscles and tendons or levers, block and tackle.

While this kind of analogical reasoning might seem curious or even simplistic today, it was widespread during the Classical era. Given the desire for evidence of vital processes, belief in the analogous relationship of plants to animals provided the impetus for devising ever more novel experiments. In these, species were isolated within distinctive surroundings in order to observe their behaviour. Moreover, in that the highest form of living creature was obviously the creator of *all* tools, no less those used by scientists, the novelty or effectiveness of experimental devices was equally a means of assessing the resourcefulness of their inventors. Similarly, in the following century widespread belief in the vital processes coursing through the house and garden made the Victorian home an experiment of sorts, a means for balancing the needs of plant, animal and human occupants and for assessing the ingenuity of householders.

THE ECONOMY OF MACHINES

Classical scientists were little aware of the various ways living beings might relate to their surroundings. Generally, they failed to question whether or not it was a matter of mutual interdependence or the adaptation of the one to the other. It mattered little whether or not innate responses or learned perceptions were involved; these are the kinds of questions posed by the environment today. During the seventeenth and eighteenth centuries, though, studies of plant and animal physiology were influenced by the mechanical or physical sciences rather than the kinds of chemical and molecular analysis that today make the organic world a very complex entity indeed. Then, there was little concern to determine what earth, air and water really were in terms of their composition. There was little thought given as to how nutrition and respiration could be described in terms other than the straightforward absorption, accumulation and expulsion of certain quantities of matter. Writing for his times, the English chemist and naturalist, Stephen Hales (1727) declared:

And since we are assured that the all wise Creator has observed the most exact proportions of *number, weight and measure,* in the make of all things; the most likely way therefore, to get any insight into the nature of those parts of the creation, which come within our observation, must in all reason be to number, weigh and measure.[4]

Given the belief that certain basic mechanical actions governed plant and animal life, two paths were opened for investigation. On the one hand, vegetalists sought means to reveal the corresponding parts of plants and animals that fulfilled similar purposes. On the other, and allowing for the appearance of different structures, they sought to demonstrate the existence of similar, continuous cycles for the assimilation and dissimilation of matter. Regarding the causes of plant mechanics, two further possibilities arose. These suggested that either actions like nutrition or respiration arose from within the plant itself or that they were the effect of some outside force. Observations were then guided by the determination of the one or the other.

These possibilities led to some interesting and varied conclusions regarding the way that living things worked. The French physicist Edmé Mariotte (1620–84), for instance, concluded that the veins in plants were responsible for forcing their juices to flow. On the other hand, his contemporary, the architect Claude Perrault, declared that the action of wind on branches, by causing compression in the veins, forced a circulatory motion much like a mechanical pump. Yet another physical scientist with an interest in the vegetable kingdom, the anatomist Marcello Malpighi (1628–94), argued for the effect of sunlight on leaves. Consequently, he described plant structure as though he were describing the components of a boiler or distillery:

I believe it probable that Nature intended that leaves serve the following purpose, namely, to allow the nutritive juice flowing from the fibers of the wood to be cooked in the utricles ... pulverized by the force of the sun's rays, mixed with the existing matter remaining in the utricles, allowing growth of new parts ... The change is the same as that undergone by the new nutrient absorbed by animals when it is diffused throughout the blood left in the vessels by earlier nutrition and transformed by it into a blood of uniform Nature.[5]

These writers described a force that they considered common to all living beings. It was available to each regardless of how and where they lived. It was present regardless of the diseases or parasites that might attack plant fibres or their exposure to wind or to obstacles that blocked the sun. For Malpighi, the action of 'sapification' that he observed in plants was simply a similar nutritional process to 'sanguification' in animals. The physiologist Nehemiah Grew (1641–1712), on the other hand, assumed that plants and animals were more or less identical in terms of their functioning parts. Grew located the force of nature within living beings themselves. Unlike Perrault, he believed the propulsive force governing plant life to be caused by the dilation of their cells, much like the action of heart or lungs. Equally concerned to relate causes and effects though conceiving of them differently, Hales emphasized a common, continuous cycle at work. An 'economy' governed plant and animal life, thus allowing for the differences between their two forms.[6]

This concept of an economy governing living things proved to be an important one. Part of the vocabulary of machines in the Classical era, it persisted in a very different form or rather, *forms*, in the era of modern science. It became a way of describing the complexity, sense of interconnectedness and regularity governing the environment. It was a way of describing a sense of balance between animate and inanimate matter, the adaptation of one to the other or the competition between species for limited natural resources. Applied to the circumstances of human existence, an economy of means encouraged thoughts on the adaptation of building forms to their surroundings.

NOVEL DEVICES

In 1727 Hales was spared such concerns, devising experiments to determine if plants and animals consumed similar quantities of nourishment and exuded equal amounts of waste and little more. Seemingly more straightforward than the diverse relationships studied by environmental scientists, Hales' idea of economy, however, was no more easily demonstrated than the balance of ecological factors that concern us today. Likewise, his need to fabricate novel devices to test his hypotheses would appear familiar to his nineteenth- and twentieth-century successors. By placing leafy boughs and defoliated limbs in vases of water, for instance, Hales concluded that the amount of water absorbed by plants was proportional to the amount of leaf area they displayed. By showing that plants generally took in large amounts of nutrient he concluded that sap could not possibly circulate, as this would suggest a closed cycle involving relatively fixed amounts of fluid. By fitting a glass tube containing mercury to the severed end of a branch, he further showed that the movement of sap varied. It was one of oscillation, moving sometimes upwards, then downwards (see Figure 2.2). Hales observed that in animals the heart expanded and contracted, forcing blood through the body, but that 'in vegetables, we can discover no other cause of the sap's motion, but the strong attraction of the capillary sap vessels [afflicted] by the brisk undulations and vibrations, caused by the sun's warmth, whereby the sap is carried up to the top of the tallest trees'.[7]

Through experimentation different aspects or departments of nature were effectively isolated for study. Both Malpighi and Hales, for instance, used measuring devices and glass containers to quantify the warmth of the sun and record the effect of its daily cycle: its power to 'pulverize', cook or invigorate nutrient-carrying fluids. Interested in assessing the effects of visible light rather than heat in his studies of perspiration, Jean-Etienne Guettard (1672–1732) devised a similarly novel means to confirm its influence. He obtained a glass sphere, one foot in diameter and provided with two openings. Through one opening in the side, a living branch was inserted. Into the other, at the bottom, the mouth of a bottle was inserted and firmly mated with the sphere. The bottle was then buried in the earth and acted as a condenser, a receptacle to measure the amount of moisture expelled from the foliage. The branch was sheltered from the direct sunlight while the temperature of air within the sphere was maintained at a temperature higher than that of the outside air. Guettard observed that the

Fig. 19.

shaded branch perspired less than when the sphere was placed in direct sunlight. He concluded that it was the 'immediate action of the sun's rays', not its warmth, that caused perspiration.[8]

John Hill (1716–75) further emphasized the role of sunlight, accounting for sensitive movement in terms of Newtonian mechanics.[9] He observed the different appearance of leaves from plants growing in various climates, a variation that 'may be assigned to its true cause, which is the different degree of light'. This conclusion depended not so much on vases, glass tubes and spheres, but on observations in the immense laboratory provided by the natural world itself. There is a hint in his research of the interests of nineteenth-century botanists such as John Hutton Balfour (1808–84) and efforts to describe the systemic arrangement of species within unique geographic regions and climates. Prefiguring the environmental leanings of nineteenth-century 'taxological' or systemic botanists, Hill described how in the East, leaves appear extended, though

not because of the heat, but because the light is strong: In the northern kingdoms they droop, not from cold, but because the air is less enlightened: In the rainy seasons they also droop, but it is not from the moisture, but the darkness of the weather; and in Egypt they are most raised of all, not because it never rains; but because the light is constant.[10]

Various hypotheses were tested in the preceding experiments, though the devices used to substantiate them each inserted the plant *into* nature, in fact, binding species fast on several occasions (see Figure 2.3). Imparting a certain unity to Classical science, mechanism allowed lessons learned of one kingdom or species to be applied to another or theoretical models and instruments used in the study of inanimate matter to be applied to the living.

2.2 An experiment for measuring the pressure of sap by fixing manometers filled with mercury to the cut ends of vines, in Stephen Hales, *Vegetable Staticks* (1727)

2.3 Eighteenth-century engraving of large globes, used as condensers in the manufacture of sulphuric acid in England and France, and by vegetalists in their experiments

In this regard, the study of fluids and gases assumed a particularly important role. Hales used fluid statics to explain both the movement of sap in branches as well as blood in animal arteries. The glass tube containing mercury and secured to a severed limb of a plant measured sap pressure while its oscillation allowed him to assert the correspondence between the force of the sun and that of the pulsing heart. Taking an entirely different approach to the cause of mechanics within living bodies, the engineer and agriculturist Duhamel du Monceau (1700–81) invoked gravity to explain why roots grew downwards and – forces being equal and opposite – the seed grew upwards. Observing root systems anchored in water-soaked earth, he concluded that:

Another force holds them fast and keeps them from separating from the moisture of the earth, namely, contact with the parts of the water and the adherence of one part to another; for there is no doubt that the moisture of the earth and the sap of the roots constitute a single continuous body, which is subject, like all bodies, to the laws of gravity.[11]

To determine the limits of such powers Thomas Knight (1759–1838) was particularly ingenious, devising experiments that must surely represent the vegetable scientist's equivalent to the torturer's rack. Knight sought to prove the generative force of gravity on plants by suspending its influence. In an experiment similar to that carried out by Humphry Davy (1778–1829) a few years later, he strapped seedlings to a rapidly spinning wheel, relying on centrifugal force to counter gravity, and then observed the direction of root and germ growth.[12] Here, as in several of the preceding experiments, the ability of the plant to regulate its own physical life and growth was understood in terms of a concept of vital force that pervaded the living being and its surroundings just as the moist earth and sap formed a 'single, continuous body'. If the wholeness, as well as the contiguity, of nature were to be known, the use of scientific instruments was crucial. Providing an isolated or momentary impediment to the force of nature as means of assessing its magnitude, glass tubes and spheres and spinning wheels in the laboratory prefigured the walls, windows and doors of the Victorian home.

CAUSES AND EFFECTS

When the preceding comments were made in the eighteenth century and shortly afterwards, there was no one, single theory to explain both the radiant and visible properties of light. Neither was there a theory of the cell as the fundamental building block of life to adequately explain the simultaneous operation of more than one vegetable process. Primarily a relationship of direct cause and immediate effect between fixed variables, natural forces and amounts of matter, theories of plant mechanics could not easily accommodate multiple forces and simultaneous plant responses such as one might now associate with the complexities of the environment. Guettard, for instance, was led to explain the whiteness of plants grown in a darkened room in terms of the absence of perspiration.[13] The concept of photosynthesis, being a chemical process involving the transformation of compounds, was not yet proposed to account for both perspiration and the presence of green pigment. Nor had the idea of chemical compounds and their forms of interaction been articulated.

The point here is not simply that the subsequent developments of a unified wave theory articulating both radiant and visible properties of sunshine, cell theory and the concept of photosynthesis were obvious improvements on Classical science per se. Rather, it must be emphasized that during the seventeenth and eighteenth centuries a particular and, with hindsight, largely mistaken view of nature prevailed. It was one that was virtually unaware of the kinds of environmental factors that only relatively recently have become objects of study. Glass tubes and spheres may have enclosed a small portion of the natural world and proved useful in revealing some small extent of its workings, but they did not provide a space for plants to *live*, complete with all the requisites for respiration, perspiration, nutrition and reproduction. These devices did not provide such a space, for the theories supporting them did not require one. They did not describe life in terms of various forms of economy between living beings and their surroundings. If a rudimentary notion of environmental influence was evinced by studies of the effect of wind on branches, sunlight on leaves or gravity on seeds, its force was simple and direct. Its force emanated within, around and through the living machine. In terms of how they were meant to work, the vegetalist's tubes and spheres removed their occupants from space and time altogether.

Were human beings similarly situated in relation to nature as plants? Not entirely, for while their bodies may have behaved similarly at the most basic level of existence, humans were obviously aware of their sensitivity to pressures and heat. They were likewise aware of the limitations of their tools for measuring these variables. There was an obvious element of abstraction in the preceding investigations of which scientists were all too conscious. In measuring the gases of perspiration, for instance, Guettard found Hales' methods inadequate for his purposes. By using a single container Hales permitted a complete cycle of evaporation and condensation to occur, thus allowing exhumed water to mix with that available for absorption. This condition was 'too close to the conditions prevailing in Nature'.[14] Hales and his colleagues grappled not only with the skills and limitations of the glassblower's art to form their experimental devices, but with their obvious artificiality. The artifice of glass tubes and spheres was known by the minor portion of the natural world they enclosed. Hence, they were inadequate to reproduce its order exactly as it appeared. If these devices prefigure the containment of nature in the nineteenth-century glasshouse, the house and garden, the experimentalist's concerns for artifice prefigure warnings by writers like John Loudon and William Robinson to shun the unnatural at home.

Spaces of near transparency

Along with interests in fluids and gases, statics and gravity, studies of light proved particularly effective during the Classical period. They were equally useful for articulating the wholeness and contiguity of nature and for relating the workings of plants and animals to their surroundings. Perhaps this was so for the most obvious of reasons. It did not take a vegetable scientist or philosopher to confirm that of all the factors which might account for growth and vision, light was truly *everywhere*. Between Locke's portrayal of the camera obscura and efforts to redesign the forcing-house, light was seen as a key aspect of a regularly performing universe, represented simply as parallel rays of energy that followed direct, predictable paths. At the same time, the occupants of these enclosures were portrayed as little more than points of focus in nature's immense, variably illuminated machinery. Describing how the one worked within the other by drawing rays of light on paper and calculating angles of reflection and refraction, the philosopher and horticulturalist alike imagined spaces within both devices that were equally abstract and diagrammatic.

LOUDON'S GLASSHOUSES

The term 'forcing-house' was one of a number commonly applied in traditional horticulture, along with the forcing-bed, -field, -frame, -glass, -ground and -pit, to describe enclosures in which vegetable growth was accelerated by artificial means, primarily heating. These structures were described by George Tod in 1812, as he noted that the attention paid to the 'forcing of plants, flowers, and fruits, for the table' had much 'increased the demand for Horticultural Buildings of every description'.[15] By isolating sunlight as the primary force responsible for plant growth, differences between species were established in terms of degree. This drew attention to the relative transparency of different types of forcing-houses, the angle of their sashes or the light requirements of the plants within them. Accordingly, as their names suggest, the significance of the palm stove, the peach house or pinery was that they each were useful in 'cooking' or stimulating a particular type of plant to early maturity, whether they were palms, peaches or pineapples (see Plate II).

The behaviour of light is clearly a concern behind George Mackenzie's recommendation in 1815 to make the form of the glasshouse 'parallel to the vaulted surface of the heavens or to the plane of the sun's orbit'[16] (see Figure 2.4). With knowledge of recent scientific developments and through his own experiments, Loudon's efforts likewise confirmed the sun's prominent role in plant physiology. Drawing on works like Herman Boerhaave's *Elementa Chemiae* (1732), Carl Linnaeus's *Amaenitates* (1745), and Adanson's *Familles des Plantes* (1763), he established a precedent for considering the placement and orientation of glass openings in forcing-houses. What had once been scientific curiosities like the refraction of sunlight and the spectral band became reconsidered in view of the horticulturalist's practical concerns.[17]

Longstanding difficulties associated with manufacturing telescope and eyeglass lenses so that they should focus light properly and limit chromatic distortion were paralleled by a number of problems for the forcing-house builder. Seeking to reconcile their forms with natural laws by not merely mimicking the lucidity of the earth's atmosphere, but by reproducing it inside, Loudon confirmed in 1822 that:

The most perfect form of a hot-house is indisputably that of a glazed semi-globe.

2.4 Elevation and plan of a curvilinear forcing-house, in George Mackenzie, 'On the Form which the Glass of a Forcing-House Ought to Have, in Order to Receive the Greatest Possible Quantity of Rays from the Sun' (1815), in *Transactions of the Horticultural Society of London* (1817)

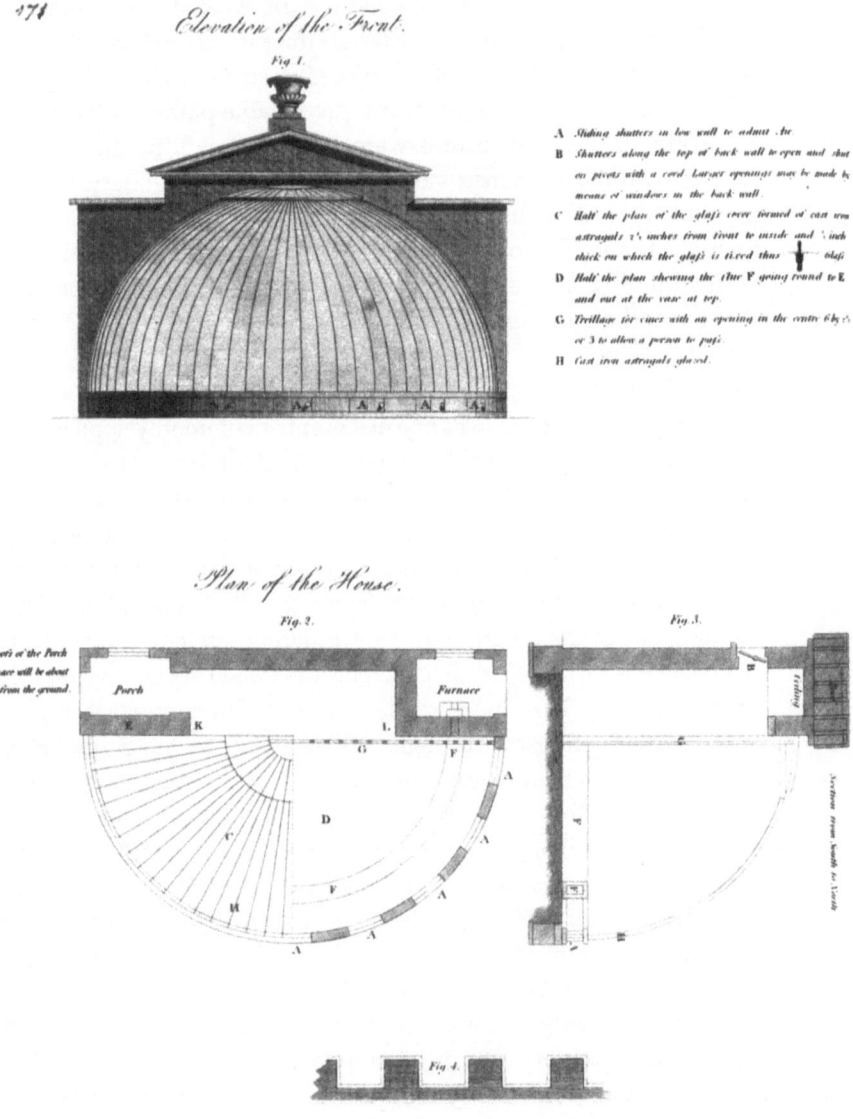

Here plants, *as far as respects light* [author's emphasis], would be nearly in the same situation as if in the open air; and art, as already observed, can add heat, and all the other agents of vegetation, nearly to perfection.[18]

Loudon's early forcing-houses were sheathed in countless small panes to form surfaces that were *more or less* curvilinear. The resulting, faceted appearance of his structures, coupled with impurities in the glass itself, were imperfections to be remedied. With the repeal of crown tax in 1845 and improvements in the manufacture of sheet glass allowing for ever larger and more chromatically pure panes, it became easier by mid-century to construct spherical forms. However, the need for them was not as great as it was once thought to be, given developments in plant physiology by that time.

In terms of how they worked, the interiors of Loudon's designs were somewhat two-dimensional in the sense that their optical performance outweighed other factors potentially affecting plant growth. Nonetheless, the forms of his forcing-houses drew attention to the importance of functional or purposefully designed structures by situating the enclosure of space within an overall economy of building. As a consequence space was itself made an object of study and technological manipulation. Mackenzie had implied as much when describing the 'Economical Hot-House' where there was

a limit prescribed to the size of a house glazed according to this (semi-spherical) figure, both by utility and convenience. A radius of 15 feet appears to me best suited to a house of this form. Anything less than this would be confined, and anything greater would render necessary an inconvenient height. Should it be desired to make the length of the house more than thirty feet, a spheroidal form may be resorted to with advantage.[19]

A concern for space was behind various proposals at the time for improving the furnaces and flues of horticultural buildings to minimize wasted heat and fuel due to poor combustion, for improving masonry walls to enhance heat retention, and ventilation systems to limit heat loss. These measures were prefaced by repeated criticisms of the opacity of traditional, shed-like hot-houses.[20]

THE CAMERA OBSCURA

The improvement of the forcing-house according to new ideas of vegetable science was preceded by John Locke's thoughts on the camera obscura. Giving form to Linnaeus's aphorism, whereas plants were meant to grow and live in the forcing-house, animals – specifically the animal part of humankind – were meant to live and feel, though not reflect upon their life or feelings, in the philosopher's small chamber (see Figure 2.5). In his *Essay* Locke wrote that the faculty of reason was like a darkened room:

External and internal sensations are the only passages that I can find of knowledge to the understanding. These alone, as far as I can discover, are the windows by which light is let into this dark room. For, methinks, the understanding is not unlike a closet wholly shut from light, with only some little opening left ... to let in external visible resemblances, or some idea of things without; would the pictures coming into such a dark room but stay there and lie so orderly as to be found upon occasion it would very much resemble the understanding of a man.[21]

In theory, by representing light as a coercive force, both the curvilinear forcing-house and the camera obscura ascribed the influence of nature to this single phenomenon. The first worked to dispel the darkness in order to enhance the exposure of plants to the sun's light and heat. The second relied on the image of a mind as a *tabula rasa* or frame awaiting a picture, positioned in the darkness and illuminated by sensations. Whereas the camera obscura re-presented the exterior world to human consciousness upon a planar surface as rays of light diverged from its narrow opening, the curved surface

of the forcing-house focused rays of sunlight upon species within. Both the curvilinear forcing-house as well as the camera obscura set apart an exterior realm of natural forces from an interior domain of physiological effects and sensory impressions. Nature, though, being the sum total of all possible forces, traversed both inside and out, belying any true distinction between inside and out.

In practical terms, however, the two devices imposed just such a distinction by creating an enclosure removed from the conditions prevailing in nature. They were a medium for relating interior spaces to exterior ones, both in terms of imposing a physical discontinuity between inside and out and by drawing attention to their obvious artificiality. By presenting a narrowed view of the outdoors, the camera obscura was no more successful at representing understanding than the curvilinear forcing-house would prove itself capable of containing the environment. This was because both devices enclosed spaces that were, in fact, obviously disconnected from the spontaneous arrangements of the natural world. In the camera obscura, for instance, the relatively small size of the resemblances cast upon its interior made projections difficult to see. It offered little opportunity for appreciating visible and ambient qualities attending a normal view outdoors. It made it difficult, for instance, to appreciate three-dimensional qualities that movement around an object or through a landscape affords. Moreover, the distorting effect of lenses and the inability of apertures to accurately reproduce peripheral vision were difficulties to be overcome technically if the device was to actually present as clear an image of the world as it was intended to communicate clarity of human understanding. Over the years efforts were made, in fact, to improve the camera obscura, along with other optical devices like microscopes, telescopes and periscopes, according to a sense of just what normal vision should be.[22] It is difficult to ascertain whether or not a sense of 'normal' vision preceded the invention of such devices or whether the opposite was the case. This latter possibility raises a related issue worth considering, involving the pivotal role played by technology in positioning living species as regards their surroundings.

Confined by thoughts on physiology and perception that were soon to be overturned, the plants within the curvilinear forcing-house and the occupants of the camera obscura were positioned by science and technology in a similar way. Basically, they remained static, motionless, located in a fixed position and acted upon by nature, sunlight and sensory impressions. In the forcing-house, plants experienced the effect of heat in much the same way as visible light since, in fact, heat and light were conceived similarly as manifestations of 'matter moving in the space between us and the heavenly bodies'.[23] Inside there was no room for the idea of heliotropism: the ability of plants to position themselves to maximize the benefit of sunlight. Though there were attempts throughout the seventeenth and eighteenth centuries to explain this phenomenon, each was constrained by a similarly deterministic framework of causes and effects. One explanation regarded plant sensitivity and movement as analogous to human feelings and actions. Thomas Knight,

2.5 Robert Hooke, 'Instrument of Use to Take the Draught, or Picture of Any Thing', detailed in a paper communicated to the Royal Society, 19 December 1694, *Philosophical Experiments and Observations of the Late Eminent Dr. Robert Hooke* (1726). Hooke's device was more akin to a camera lucida than to a camera obscura, though it served a similar purpose to the latter, namely to reproduced scenes 'truthfully'. Hooke sought to help navigators and travellers determine where they were by creating a device for depicting the coastlines or countrysides sides they encountered

noting that leaves always position themselves so that light falls squarely on their surface – a phenomenon called diaphototropism – conducted an experiment in 1807 whereby he twisted vine leaves into a variety of positions and placed all manner of obstacles between them and direct sunlight. Despite having 'put obstacles in its way on which ever side it attempted to escape', he nonetheless failed to find 'anything like sensation or intellect in plants'.[24]

Without an understanding of cellular chemistry and no reasonable theory like evolution to suggest otherwise, the exact cause or justification for the responses of plants to nature was seen to be ordained by God, while the vital force responsible for their movements transcended the plant's machine-like character. Likewise, the thoughtful activity of man was seen as distant from the apparatus of observation and sensory experience. The source of self-consciousness lies somewhere beyond the camera obscura, for as Locke wrote: 'The Understanding, like the Eye ... takes no notice of itself.'[25]

Places of impenetrable darkness

Historically, there were other places, actually occupied for certain lengths of time – notably the greater portion of a working life – that provided opportunities for observing the reactions of living beings to particular environs. They contained humans on the whole, rather than plants like the forcing-house, and were less about ideas and resemblances than labour, unlike Locke's camera obscura. Spaces aboard Britain's ships and in its mines, its foundries and factories at various times in the past presented certain features in common with these devices. They were almost always confined and notably deficient in one or the other requisites of life. In being so, they drew philosophers and technicians to consider how human vitality depended on certain things, like light and air. These spaces provided laboratories, as it were, as important as any horticultural building or philosophical construction might be, in which ideas about the workings of mind and body were derived from observations of situations in which human respiration, perspiration and nutrition suffered. They were places where ideas about what a proper environment *should* be like were formed. They contributed to a growing body of evidence for nineteenth-century scientists to discern the facts of human life.

Alain Corbin tells us that ships, along with schools and prisons, offered instances in which the effects of crowded spaces could be experienced long before they became an issue for nineteenth-century social reformers and urban planners. These were instrumental in the formulation of both hygienic science and psychology and furthered thoughts on the nature of the senses of smell and personal space. Captain Cook's *Endeavour* is particularly interesting for it was depicted as 'the first hygienic city in miniature'.[26] Ten years after publishing his *Statical Essays* in which Stephen Hales applied investigations into the composition and behaviour of air to the study of the circulation fluids in plants and animals, the vegetalist described various devices, air-pumps and ventilators for exchanging the noxious vapours of ships, mines, gaols, hospitals and workhouses with plenty of fresh air:

> that genuine Cordial of Life: For that wonderful Fluid the Air, which, by infinite Combination with natural bodies, produces surprising Effects, as it is on the one Hand when pure, the chief Nourisher and Preserver of the Life of Animals and Vegetables; so, when foul and putrid, it is the great Principle of their Destruction.[27]

The long history of mining and ore extraction in the British Isles and continental Europe was accompanied by similar efforts to ameliorate the harmful consequences of life in close quarters. Books described ingenious machines for delivering fresh air underground and for removing 'heavy and pestilential vapours'. Georgius Agricola's scholarly work on mining and metallurgy, *De Re Metallica* (1556), illustrates many of these (see Figure 2.6). It exploits developments in sciences of the earth, minerals and engineering not to transform elemental nature or to return to the alchemist's 'primordial chaos', but to keep those that toiled under the earth alive.[28]

By the early decades of the nineteenth century, the impurity of air in many workplaces led to investigations establishing the longevity rates of practitioners of various occupations. There arose the general conclusion that life, its duration and quality, was not simply preordained or a matter of fate. Rather, it could be improved through technical means.[29] In addition to treatises on the reform of factories and workshops, works calling for the improvement of housing of rural labourers provided a wealth of detailed study. They detailed the consequences for health when cottages were situated in ill-drained districts, close to sources of fogs and malaria or near 'rotting places' in the fields.[30] Some of these concerns had been raised in previous eras and during times of plague, though they were to acquire an empirical, scientific cast and heightened social urgency. Plague, a term which had once encompassed various illnesses of unknown origin, was replaced by 'fevers' arising in more identifiable, institutional confines in eighteenth-century Britain. Hence, there was 'gaol fever' in prisons, 'army fever' afflicting the military and 'hospital fever' in hospitals.[31]

The cramped, fetid and disease-ridden spaces of these various places could conceivably be described as counter-images of the rationally planned city with its modern, sanitized compartments. Little more will be added here, except to suggest that they are worthy of further study in their own historical contexts and not simply for revealing the cumulative and progressive character of Western science in overcoming the deficiencies of pre-industrial spaces. It appears rather more interesting to consider how these spaces, vanquished or largely regulated in many parts of the world though clearly not all, were likewise means for articulating environmental concerns. This occurred on a more significant scale than either the camera obscura or the forcing-house. The intimate contact with earth, air and water in confined spaces and their contamination by human activity over the course of Britain's long industrial history provided opportunities for some ideas about nature to be tested and rejected and for others to endure.

2.6 Machine made from bellows for providing fresh air and removing 'heavy and pestilential vapours' from mine shafts, in Georgius Agricola, *De Re Metallica* (1556)

Putting these thoughts aside, it remains to be seen how the demise of a primarily mechanical view of plant and animal behaviour, reason and human experience represented by the curvilinear forcing-house and the philosopher's camera obscura was played out in the nineteenth century. At the time of Mackenzie's investigations and Loudon's glasshouses, the concepts of the organism and its adaptation to something called the environment and the inhabitant and its conscious regard for its surroundings were, as yet, little explored possibilities. The possibility that an adaptive power might precede physiological response in plants and humans, just as thought precedes and organizes experience for humankind, was not admitted. Loudon's plants remained fixed in place, receptive perhaps to rays of light drawn on paper, but little more, just like the prisoner in Locke's dark room.

François Delaporte observed how the vegetalist's practice of removing potted plants to glasshouses and enclosed rooms created controlled conditions for observing the natural world, though in a much narrowed sense. The practice eliminated seemingly extraneous factors such as wind and insects to enhance the results of experiments. This flattening of the world occurred in devices like the camera obscura and in places of commerce and industry in centuries past. Similarly, it was the influence of factors *outside* the glasshouse as well as the Victorian house and garden that proved most thought-provoking. These spaces encouraged visitors to the botanical garden and family members in the home to situate themselves within their animate and inanimate surroundings. It called upon them to be more consciously aware of the broader landscape of organic existence, though no particular certainty was afforded them. Rather, by falling outside the limits of human reason, science and experimentation and by resisting efforts to contain them, these myriad organic factors and relations that seemingly could not be contained drew attention to the environment.

Habitats and ways of life

Given the influence of the science of vegetality on the curvilinear forcing-house, there was no qualitative distinction to be made between the external forces acting upon the structure and the reactions of plants within. The impact of sunlight upon leaves produced an automatic and immediate response. The forms and surfaces of botanical specimens were mere registers of solar activity, of light and heat. Nevertheless, in drawing attention to the sun's light and heat, the containment of space within the forcing-house made it obvious when plants did not 'work' as they were supposed to. Similarly, the scientist's experiments could reveal the inadequacies of built form and structure. Conflicts like those arising between Thomas Knight's sun-seeking grape leaves and a concept of nature that limited possibilities for a physiological explanation of sensitive movement directed efforts to uncover more fundamental adaptive processes within the structure of organisms themselves. Scientific vision, in other words, could no longer rest upon the surfaces of things or rely on machines to describe the function or purpose of living bodies. Observation and experiment were directed towards a more fundamental level of organic existence, namely the cell and its chemical operations. There, the Classical discourse of vegetality disappeared.[32]

In studies of the workings of human experience, it took the observation of such curious phenomena as post-retinal images for philosophers to abandon the camera obscura as a model of understanding. It was soon accepted as given that thought organized sensory experience to a large extent, not the other way around. Whereas Locke remained silent on the physiological workings of the mind, the formulation of the 'Bell–Magendie Law' in 1822 stressed the active role of the brain in organizing consciousness. Dispelling the darkness and squalid conditions of Britain's ships, mines and factories, the rationalization of life and labour over the course of the nineteenth century was accompanied by a new sensitivity to the interaction of human physiology and psychology. It was accompanied by a belief that the condition of the inmates of these spaces could be improved along with their surroundings.

The emergence of biology provided a powerful justification for the growing belief that the relationship of a living being to its surroundings was not simply direct or mechanical, but was assigned a purpose within the overall logic of a species' evolution. The historian of science Alan Morton observed that

the real break with the past in biology was not sudden, but it began unmistakably during the eighteenth century with the dawning recognition of the possibility, and then the probability, that living organisms had not reproduced themselves unchanged since some moment of divine creation, but had undergone a development in time, in which simpler organisms gave rise to more complex ones, fitted for many habitats and ways of life.[33]

Development involved immediate, sensitive responses. For plants these included heliotropism and geotropism; for animals, reflex actions. Growth and reproduction incorporated long-term adaptive strategies such as the transformation of species and their phylogenetic regularity. This is a way of describing the 'race history' of an organism or species as opposed to the development of individual beings. Diverting attention from the narrow study of nutrition, respiration or other metabolic processes, the actions of adaptation and evolution were prescribed by an economy of environmental resources and forms of consumption. As Charles Darwin made clear, these required the survival of the fittest species.

Within this diverse, though increasingly legible, field of inquiry, biology was guided by methods that sought to locate the level of adaptive responses within living beings. This had the effect of isolating physiological responses to light, heat or air and specific environmental stimuli in order to determine what they really were. For the scientist it was no longer adequate simply to compare the corresponding parts of plants and animals or to demonstrate the similarity of their vital processes. Rather, it was important to ascertain just where these processes occurred and just how matter was transformed in the body. It was no longer adequate to assess the action of wind on branches or movements within plant cells, without tracing the species' evolutionary history. It was important to acknowledge, along with other facts of life, that the distribution of limbs and the pliability of trunks help trees withstand the force of gales and other, equally unfortunate circumstances.

CELLS, ELEMENTS AND LIGHT WAVES

Some key developments that contributed to a new basis for the study of plant and animal physiology are worth noting. A credible theory of the cell, for instance, capable of explaining both photosynthesis and the sensitive response of plants to their surroundings, was proposed in 1839 by German botanist Matthias Schleiden and zoologist Theodor Schwann. The composition of cells and the description of protoplasm as the site of sensitive movement – the 'vehicle of irritation' discerned by Knight – was established during the years 1844 to 1855.[34] Whereas the conclusions of eighteenth-century experimentalists such as Hales and Guettard were guided by the state of knowledge regarding the composition and behaviour of matter, the description of photosynthesis depended upon the eventual demise of the theory of phlogiston – a long-lived concept involving a principle of combustion that produced heat in both animate and inanimate bodies and, in the former, vitality. This belief was finally discredited by Antoine Lavoisier following his chemical experiments conducted between 1786 and 1789. Prior to that time, investigators such as Joseph Priestley, Jan Ingenhousz and Jean Senebier had to rely upon a variety of often confusing and qualitative terms to describe the varieties of the material that seemed to enable combustion: the substance now known as 'gas'.[35] Experiments, particularly those carried out by Priestley from 1767 to 1786 into the constitution of this substance, led to the creation of a standardized chemical nomenclature. This was furthered by the elaboration of Dalton's atomic theory in 1808, which was eventually extended to the analysis of other substances, such as soil and water, essential to life processes and to the articulation of those processes and their forms of interaction.[36] Foreshadowing a new domain of environmental inquiry drawing on the complex relationships appearing between animate and inanimate matter, John Pringle wrote on his presentation to Priestley of the Copley Medal of the Royal Society in 1773:

From these discoveries we are assured that no vegetable grows in vain; but that from the oak of the forest to the grass of the field, every individual plant is serviceable to mankind; if not always directly by some private virtue, yet making a part of the whole which purifies and cleanses our atmosphere. In this the friendly rose and deadly nightshade co-operate; nor is the herbage, nor the woods, that flourish in the most remote and inaccessible regions unprofitable to us, nor we to them, considering how constantly the winds convey to them our vitiated air for our relief and for their nourishment.[37]

As the power of light figures prominently in this study so far and will continue to do so, new ideas regarding its composition and behaviour are worth noting. Many early developments in optics from 1600 to 1800 were guided entirely by the observation of light's visible aspects.[38] Isaac Newton had observed the dispersion of light and, with Robert Hooke, the colours of thin plates. This led to the description of interference in terms of Newton's 'rings' – the curious pattern of reflected and refracted light caused by its passage through a prism – though the first quantitative explanation of this phenomenon was not made until a century later in 1802 by Thomas Young. Polarization by reflection was discerned in 1808 and fluorescence in 1845. David Brewster elaborated his findings on the reflection and refraction of polarized light in 1818 according to the Newtonian model whereby light was thought to consist of rays of 'corpuscles', or minute particles of matter. These should not be confused with atoms or molecules, which were unknown at the time. Observable phenomena supported both corpuscular and wave theories of light, though Augustin Fresnel's analysis of polarization in 1823 was ultimately to establish light waves as a theoretical model able to accommodate its peculiarly composite nature.

What is important to consider is that these developments were not necessarily advances on the theories of 'phlogistonated' air or corpuscular light rays per se or that they led to the refutation of vegetality with its basis in analogical reasoning. Rather, these new chemical and physical methods allowed observations of diverse phenomena found in nature and comprising an environment to be related to one another in a reliable or calculable manner. Chemistry recast the human being's relationship to the vegetable kingdom by describing the basic elements of the one that could be used, consumed and transformed by the other. In 1813 Humphry Davy (1778–1829) noted the progress of a new kind of 'agricultural chemistry':

If the organs of plants be submitted to chemical analysis, it is found that their almost infinite diversity of form depends upon different arrangements and combinations of a very few of the elements [...] and according to the manner in which these elements are disposed, arise the different properties of the products of vegetation, whether employed as food, or for other purposes and wants of life.[39]

In the study of animal physiology, François Magendie (1808) rejected the simple analogies formed by his predecessors between mechanical hydraulics and the circulation of fluids in living beings.[40] Following various experiments in 1822, Magendie also established the functional division between sensory and motor nerves, a conclusion similarly reached on anatomical grounds by Charles Bell. The significance of such an assertion was not only that the components of understanding could be analysed physiologically rather than philosophically, but that study could proceed in a scientifically rigorous manner. Subsequent elaborations of their theories over the next half-century were to allow physiology and psychology to be related experimentally.[41]

SENSE AND INTELLECT

Following the preceding discoveries, investigations into the character of adaptive responses in animals were focused on the role of the brain. As an organ, it was determined to be affected by growth and age and hindered by disease. As the mind or seat of consciousness, it operated to transform experience into understanding. Subsequently, Alexander Bain was to show in 1855 that knowledge was the product of experience consequent to action and not merely a complex ensemble of ideas that one might define epistemologically. An implication of his work was the link that could be established between the sensation of such ambient qualities as light, warmth and sound and a living being's occupation of space and movement through it. No longer fixed in place like a plant in a forcing-house or the spectator in the camera obscura, the idea of the inhabitant was cast in terms of a being that exploited a full range of senses and bodily movements and interpreted their surroundings accordingly as a principal means of adaptation. Likewise environmental qualities became the product of temporal existence and not the character intrinsic to objects or spaces for, as Robert Young related:

Once an important role had been found for motion in learning, interest in the topic naturally spread to behaviour itself. The new biological context for associationism (which was beginning to manifest itself in [Herbert] Spencer's work on psychology which appeared in the same year as Bain's first book and in Charles Darwin's work which appeared simultaneously with Bain's second book) greatly advanced the development of an interest in behaviour and the adaptation of organisms to their environments.[42]

A belief in the adaptive role of the mind recast the fundamental distinction believed to have existed between plants and animals. Biology supplanted vegetality in describing the processes shared by both classes of beings, setting the limits of botany on the one hand and of zoology on the other. These studies provided a measure of organic reality against which consciousness and its influence on perception and human cognition could be assessed. Seven years after failing to find proof of 'anything like sensation or intellect in plants', Knight differentiated between these two attributes, pointing to the first as a means of confirming the presence of organic truths. In the following passage by Knight, sensation is not simply a means of attributing feelings to animals or waging a theological battle over the rights of humankind to dominate lesser creatures, but foreshadows a domain of environmental concerns linking all species:

It is true, that plants do not appear to possess sensation in the ordinary sense of that term, as it is applied to animals; but nature, in forming its whole *organic* creation, seems to have preceded so much by substitutions and additions, that simple sensation, in its strict and limited sense, *abstracted from all powers of perception* [author's emphasis], may not improbably be as widely diffused as organization itself; and animal and vegetable life may be, in consequence, susceptible of similar injuries from similar external causes.[43]

Following a period of shifting theories from which the life sciences emerged, it became impossible to describe the spaces of glasshouses and darkened rooms without acknowledging the experience and ultimately the 'psychology' of their human inhabitants. It became impossible to ignore the responsibility of their inventors or designers for improving the quality of life therein. It became impossible, in other words, to study the life forms in various enclosed or confined places without learning something of the spaces required for human vitality and well-being as well. The 'sleep of plants', as John Hill (1757) described the curiosity of vegetable sensation and the closure of leaves at night or upon touch, took on new meaning as physiological behaviour became distinguishable from human desires for a comfortable life.[44] Accordingly, it is necessary to consider more carefully the development of the glasshouse as a structure conceived and built *by* humans *for* plants in the light of terms set by biology and, ultimately, environmental and human sciences.

For the biologist concerned with physiological processes, the sunlit, heated and humidified space of the glasshouse affected the spectator within the conservatory as much as the plants it contained. For the environmentalist interested in models to represent the complex organic world, glasshouses, although perhaps unable to fully accommodate or imitate nature, evince a fundamental human responsibility to care for it. For the behaviourist interested in the reaction of human beings to their surroundings, the sensibilities engendered by an experience of the conservatory reveal much about their inhabitants' social or cultural inclinations as well as their psychological make-up. More on this will be said later. For now, these observations illuminate what seems at first glance to be straightforward, practical advice. Writing in 1814, Knight warned gardeners not to allow their own desires to outweigh the needs of plants in their care:

Being fully sensible of the comforts of a warm bed in a cold night, and of fresh air in a hot day, the gardener generally treats his plants, as he would wish to be treated himself; and, consequently, though the aggregate temperature of his (forcing) house be nearly what it ought to be, its temperature during the night, relative to that of the day, is almost always much too high.[45]

The measure of life

It seems natural to value the distinctive, progressive and cumulative character of Western science when questioning the order of nature. It seems normal to admire the technologies capable of transforming it. Our wonder, however, stems from the assumption that space and time exist independently of reason. Space and time are commonly taken for granted, 'an unshakeable frame of reference inside which events and places would occur'.[46] By this, Bruno Latour drew attention to the way in which our faith in science relies upon a distinction that is commonly made between a way of knowing things and an extant world waiting to be known, somehow 'out there'. He argued to the contrary, that sciences like biology, botany and zoology exist only within networks that include institutions and distinctive practices of collecting, naming and labelling. These practices effectively locate phenomena in relation to one another, causally or chronologically. Thus, the dream of a universal taxonomy so closely associated with natural historians such as Carl Linnaeus (1707–78) and Jean Baptiste Lamarck (1744–1829) was only transformed into knowledge within the confines of Kew Gardens in London or the Jardin des Plantes in Paris. These were places where plant species were removed from their places of origin in order to be named and labelled, germinated and cultivated. The lesson of the forcing-house and the camera obscura is similar. Though subjects of fairly outmoded scientific and philosophical discourses by the end of the eighteenth century, they nonetheless revealed the causal and temporal links arising between living beings and their non-living surroundings.

The purpose of the forcing-houses proposed by Mackenzie and designed by Loudon was to regulate the forces of nature, not to reproduce them. In other words, their purpose was to keep the living machine 'wound on'. Nature was itself envisioned as an entity separate from the various technological means intended to engage it, something, somehow 'out there'.

In practice, though, the forcing-house, like Locke's darkened room and spaces conceived or made by humankind like ships, mines and factories, revealed nature to be a much more complex entity than was originally assumed. This was because they failed to 'work' as they could have. In the glasshouse the environment supplanted nature as the concepts of life and the functional interdependence of species and their surroundings assumed the transcendent status that once made nature an all-pervasive force in the forcing-house. The relationship between living kingdoms changed significantly as mechanicism was rendered untenable.

In the nineteenth century plants and animals assumed a new, common relationship to an external world – their shared environment. It was not only the source of positive effects, stimuli and morbid agencies, but was a medium for exerting powerful influences on one another. Two fundamental distinctions arose along with the sciences of life. In the first instance, a more fundamental opposition between organic and inorganic existence appeared. This had been raised by Lamarck at the end of the eighteenth century to challenge the longstanding distinction between animal and vegetable kingdoms. The organic was all that lived, grew and reproduced itself. The inorganic 'lies at the frontiers of life, the inert, the unfruitful – death'.[47] In the second instance, the living being, whether plant or animal, came to emanate within, around and through the environment. It was not only a body that made the reality of the environment apparent by thriving or dying in its surroundings. The living being also imposed itself on others. In 1774 Priestley, for instance, described 'the restoration of air, in which a candle has burnt out, by vegetation'.[48] Recalling this discovery five years later,[49] Ingenhousz was one of the first to have theorized the interdependence of animals and plants on the basis of respiratory and nutritive processes shared *between* them as opposed to a relationship based simply on analogy.[50]

The idea of the environment encompasses varied relations between organic and inorganic matter and plant, animal and human species.[51] Subsequently, the living beings that grew and reproduced themselves became differentiated from one another on the basis of fundamental functions. This involved, for instance, reproduction for plants and alimentation for animals. They became classified as to how they lived, whether deciduous or evergreen, herbivorous or carnivorous. They were sought on the basis of where they lived, whether in the desert, the tropics, or an alpine meadow. They were seen to be themselves determined by these environs. Deprived of mechanical certainty, but armed with a range of technologies for reproducing these spaces of habitation, humankind made all living things subject to their expectations and desires. Thomas Knight foreshadowed these developments, revealing a hitherto unexplored realm of environmental manipulation. It was 'an ample and unexplored field for future discovery and improvement [that] lies before us, in which Nature does not appear to have formed any limits to the success of our labours, if properly applied'.[52]

As for human involvement in the spontaneous arrangement of the organic and inorganic world, the late eighteenth-century painting *Experiment with the*

Air Pump by Joseph Wright of Derby is prophetic. The experiment depicted draws on studies into the composition and behaviour of gases carried out by Joseph Priestley at roughly the time the painting was made (see Plate II). The glass globe, both in art and the laboratory, became a transparent medium by which air – as merely one aspect of nature – was revealed through its enclosure, quantification and manipulation. The painting contributed in a small way to the popularization of scientific themes, particularly as it allowed a hitherto unrecognized realm of natural phenomena to be experienced close at hand. Capturing the moment that air is drawn out of the pump and its occupant begins to suffer, Wright's work prefigures a particular sensitivity to artificial climates and their effect on living organisms, though it is not a painting of an environment per se. The bird depicted in the air pump, a crested cockatoo, evokes a world that was little explored prior to subsequent theoretical and technical developments in the first half of the nineteenth century, though the notion of an economy of natural forces and causes and effects is there. Nature, soon to be reduced to cells, elements and light waves, will remain no less mysterious than before. The reactions of the audience in the painting, expressing both wonder and anxiety – perhaps even horror – prefigure reactions to the environment today.

The vital landscape

In 1831, following his experiments with curvilinear forcing-houses and while receiving critical acclaim for his *Encyclopaedia of Gardening*, John Loudon proposed an extraordinary new structure for the grounds of the Birmingham Botanical Horticultural Garden. In a vast iron-and-glass dome over 61 m (200 ft) in diameter, with concentric beds of plants and walks, and radiating aisles and an imposing central observation tower with viewing ramps that spiralled to the ground, the order of nature was carefully arranged for viewing by the public. It was the second of two schemes he illustrated for the Garden and the one he preferred (see Figures 3.1 and 3.2). The first, of less height though comparable monumentality, encircled a tower containing a heating apparatus and potting shed. The convenience of the gardener in this scheme apparently took precedence over the orchestration of spectacle.

Coincidentally, both proposals were followed one year later by the death of Jeremy Bentham, whom Loudon had long admired. Melanie Simo notes the similarity in form of the glasshouses and the Panopticon. In Bentham's model prison the judicious use of space and the principle of surveillance were coordinated to turn 'rogues' to a better, more virtuous life.[1] Beyond nurturing plants, the broader aim of Loudon's glasshouses was to provide a space in which viewing nature was itself a virtuous act. The great dome of Loudon's Birmingham Garden project echoed his prophetic announcement in 1817 of a new concern for both horticulturalists and architects: the challenge of environmental control.

When subsequent improvements in communicating heat, and ventilation, shall have rendered the artificial climates produced equal or superior to those which they imitate, then will such an appendage to a family seat be not less useful in a medical point of view, than elegant and luxurious as a lounge for exercise or entertainment in inclement weather. Perhaps the time may arrive when such artificial climates will not only be stocked with appropriate birds, fishes, and harmless animals, but with examples of the human species from the different countries imitated, habited in their particular costumes, and who may serve as gardeners or curators of the different productions.[2]

The space that Loudon sought to contain in his dome was a subject so

3.1 Elevation, section and plan of John Loudon's first scheme for the Birmingham Botanical Horticultural Garden, 1831, appearing in *Gardener's Magazine* (August 1832)

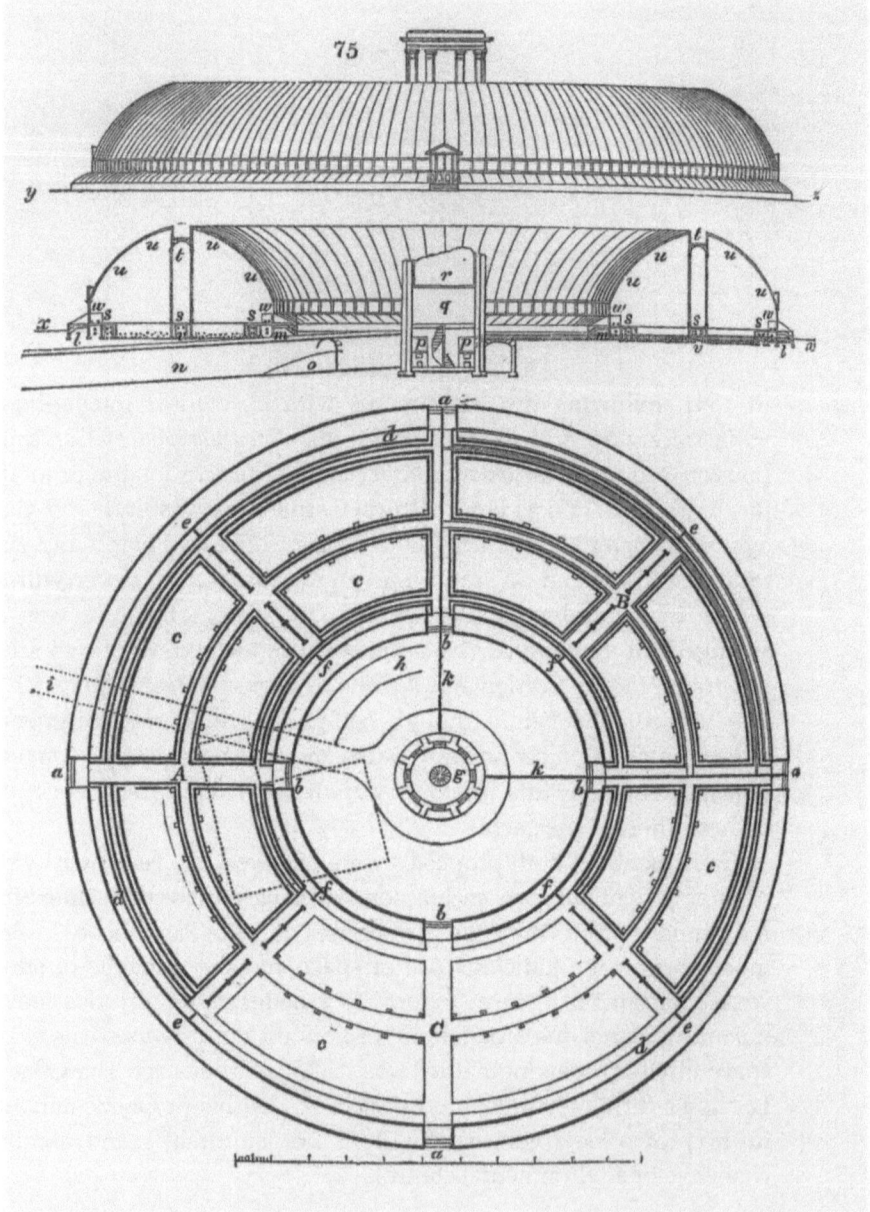

'new and strange' that he feared the ridicule of readers. Now it is an idea so familiar as to be taken for granted. It is the environment: the space of organic existence, of life with its multitude of forms and the links that bind organisms to one another and to their surroundings. In the first decades of the nineteenth century this domain had been opened only partially to scientific inquiry. It was perhaps slightly more so for imaginative, rather than experimental, exploration. Whereas a longstanding belief in the fullness of nature may have led the natural historian to distribute living species on a

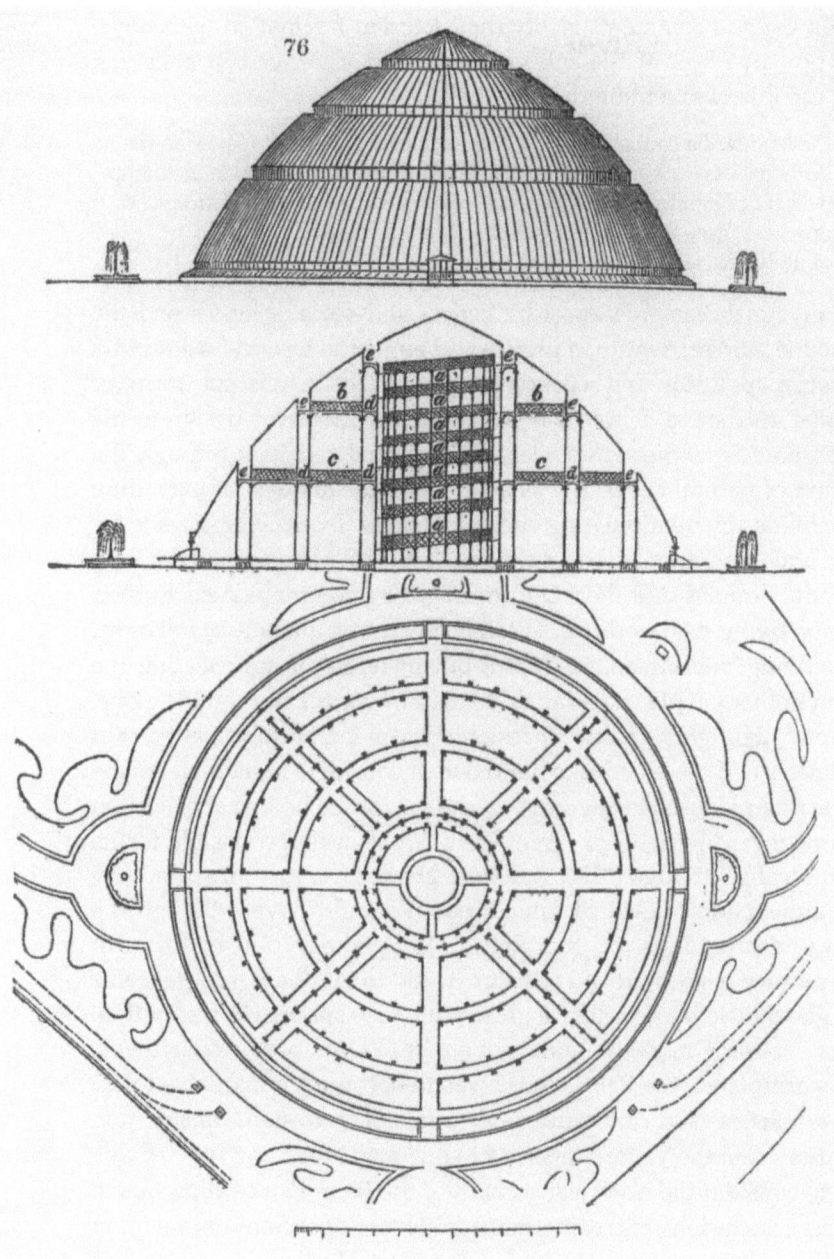

3.2 Elevation, section and plan of John Loudon's second scheme for the Birmingham Botanical Horticultural Garden, 1831, appearing in *Gardener's Magazine* (August 1832)

taxonomical table according to differences between their physical features, it was recognized that it did not necessarily follow that plants and animals were fixed in place in terms of unique situations, geographical locales and climates. By the 1830s an empirical understanding of plant and animal physiology had just begun to reveal the reality of their interdependence and the ways living beings were adapted to suit their surroundings. It was just then possible to think of artificial climates as forming an environment in the modern sense of the term. In his *Wanderings Through the Conservatories at Kew* (1857), Philip Gosse's comments typify the Victorian fascination with novel environs. They also hint at the self-satisfaction they derived in believing themselves capable of inventing them, as only

A door, and the length of a couple of yards, conducts us from the temperate to the torrid zone. Without was a pleasant genial English summer day, within is the damp and oppressive heat of Hindostan. Into what noble society are we now introduced? What regal forms are these which stand bathing their green crowns in such an atmosphere of moisture, warmth, and light?[3]

Accompanying the emergence of biological and social sciences and the project to define human identity in precise and empirical terms was a rapidly evolving social, political and economic context for a nascent form of environmental awareness. It was a field where attention was drawn to the form of glasshouses as means for elaborating social relations through the representation of natural order. It was a source of techniques for extending human capabilities by manipulating enclosed spaces. By providing a site for observation and experimentation, glasshouses raised a context for thinking about just how humans and their culture might be accommodated to their living and non-living surroundings. Thoughts were cast not only in universal terms of bettering the human condition, but in terms of appreciating the different ways other people had adapted to their own situations on the globe.

This chapter highlights circumstances accompanying Loudon's interest in artificial climates and the attempt to build one. It considers in more detail the events in the Palm House at Kew in 1867, the fate of the botanical collection and questions regarding who or more likely 'what' was responsible for its demise. Ironically, though the iron-and-glass structure inspired by Mackenzie and Loudon failed to function as it should have – though the theoretical underpinnings of the Palm House's curvilinear form in eighteenth-century science proved to be unsteady to say the least – its iconic status as a glasshouse par excellence grew with each successive restoration. Incapable of enclosing in one volume alone nature's complex arrangements when it was completed, the Palm House was to become a remarkable token of nature's wholeness and contiguity as accounts of its design, construction and celebration from the Victorian period to our own attest.

Turning to consider the historical account of the Palm House allows us to draw broader conclusions regarding the significance of the environment, in both its natural and human-made forms. These conclusions impact on our understanding of human identity, integrity and character today. By the middle of the nineteenth century glasshouses, like the houses, gardens and other contrived landscapes, drew constantly on the resources, materials and idea of nature and the unique characteristics of particular sites, locales and environs. In their leafy, sometimes carpeted, lovingly terraced confines, householders and their designers cultivated particular attitudes to nature in the form of scientific principles, all the while hoping to satisfy newly found necessities of life. Novelty was contained in the Palm House. This is evident not only in its curvilinear structure and the exotic species it sheltered, but in the revelation that it did not entirely 'work' the way it should have. Perhaps the history of the Palm House at Kew offers a clue as to why our houses often fail to work any better.

The cultivation of reason

The allure of artificial climates was in part made possible by developments in science and technology, but also by circumstances that encouraged inventiveness in identifying, then satisfying, life's necessities. Intrigued by the vital processes exhibited by living species, eighteenth-century vegetalists were led to invent novel experiments to measure their effects. From a broader perspective, appeals to science and natural law by reformers in the nineteenth century were guided by fascination with forces of exchange and consumption, trade and labour coursing within the social body. They led to thoughts on how these might be facilitated by new urban forms and spaces, acts of Parliament and other means.

Cast between the interests of horticulturalists and those of social and political theorists, the glasshouse presents a particularly interesting study. In the form of the forcing-house, it not only provided evidence of the reality of biological existence by revealing the complex processes and economies of nature. More generally, as a building type encompassing many forms, including curvilinear ones, the glasshouse enclosed a ground on which science was cultivated as the basis of professional expertise. This involved forming hypotheses about how the structures should work and drawing conclusions as to why they were less than satisfactory in the past. Science was promoted along with a language of forms, functions and appearances, means and ends, causes and effects. It was a context both for articulating problems and for distributing responsibilities for solving them. It was a means for distinguishing between the talents of those members of the professional classes: horticulturalists and landscape gardeners, architects and builders.

HOME CRISES AND SCOTTISH ENLIGHTENMENT

Writers like George Mackenzie and his contemporaries, George Tod, John Williams and Henry Home, were led to experiment with the common forcing-house not by scientific curiosity alone. Rather, they were guided by circumstances that encouraged them to turn from the comfort of a 'study lined with books' to use science for practical purposes.[4] By the end of the eighteenth century circumstances were such that fears arose in the United Kingdom of an impending agricultural crisis. Assessments of inadequate domestic yields and rising land prices, the loss of foreign markets due to the Revolution in North America and naval blockades during the Napoleonic Wars focused thoughts on the problem of feeding Britain's population more efficiently and profitably. The forcing-house was one means for doing so (see Figure 3.3). Improvements in the manufacture of iron subsequent to 1784

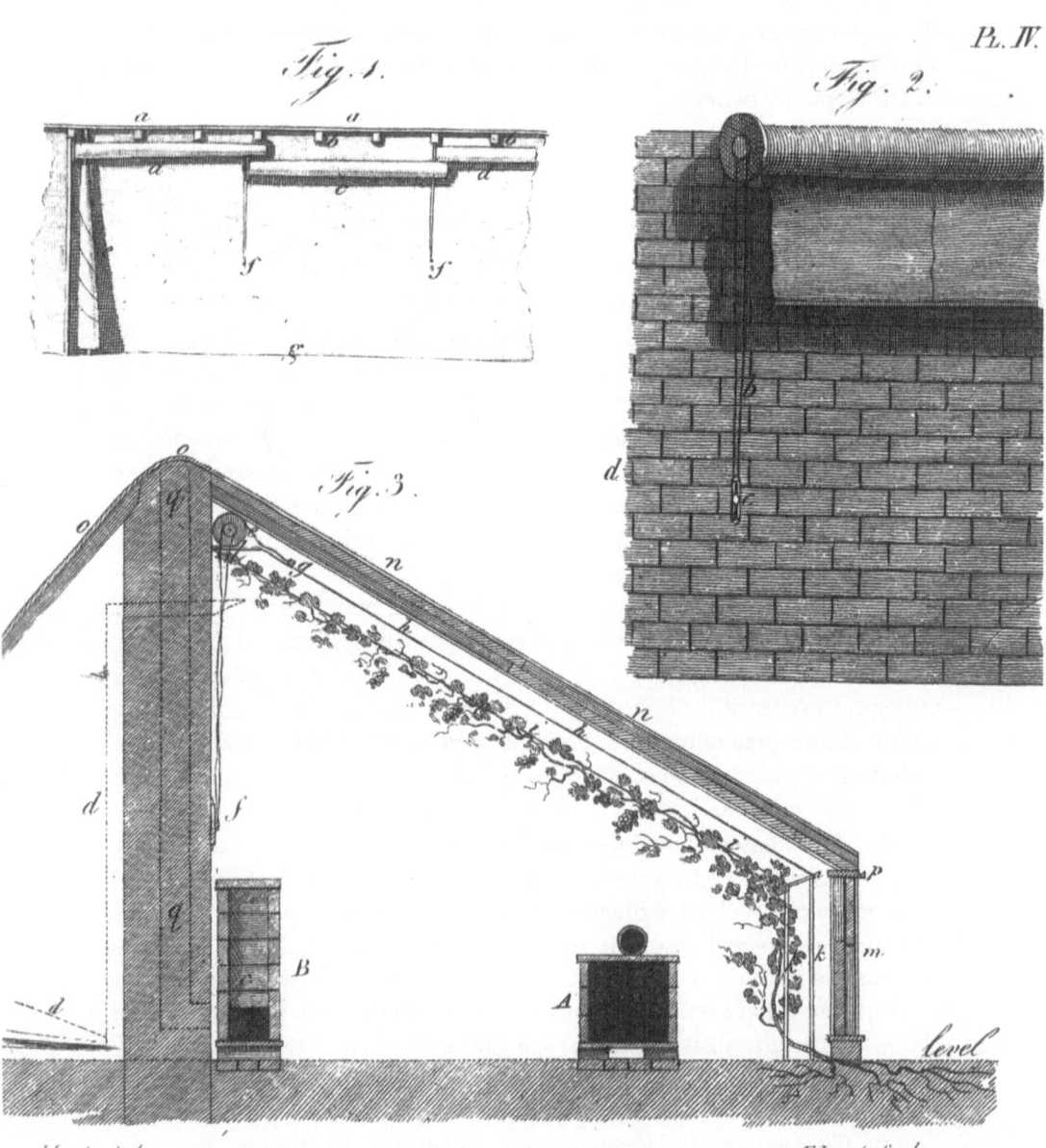

J. Loudon Del.   F. Lamb Sculp.

were coupled with anxieties over diminishing supplies of domestic timber for shipbuilding and other purposes. Improvements in glass manufacture were heralded by protests leading to the repeal of various taxes governing the use of glass in 1813 and again in 1845. These changing circumstances encouraged the use of new structural principles and materials in the fabrication of glasshouses throughout the first half of the nineteenth century.

As one sector of an increasingly legible though ever-complex economy, the state of agricultural production in the final decades of the eighteenth century added urgency to the efforts to use state power in order to consolidate national commercial interests. These were initiated by the Georgian monarchy and aimed to promote manufacturing through nationwide strategies for improving domestic production of all kinds of goods.[5] In a letter written in 1787 to Adam Smith, Bentham described the dependence of manufacturing on agricultural production, contradicting commonly held opinion that the opposite was the case. Outlining the benefits that an increase in both promised for the public good, he prefaced his comments with a statement of biological necessity. He claimed that 'When men have got more food than they want, they will be able and willing to give part of it for finer clothes and finer furniture.'[6] The invention of a better glasshouse for agricultural purposes was not only a response to newly formulated needs of plants for sunlight and the prospect of famine. It was also a reaction to an expanded sphere of natural resources to be exploited and human energies to be realized. The glasshouse raised concerns of both British industry and the 'industriousness' and inventiveness of the British character.

Scotland provided fertile terrain for new thoughts about science and its usefulness, as works so far cited by Scots Mackenzie, Loudon and Fergusson suggest. While union with England in 1707 served to depress the Scottish economy by creating direct competition with England, the university at Edinburgh became increasingly instrumental in applying science to the improvement of Scottish industry, providing a source of ideas that were carried south.[7] At the time of Loudon's early writings, Edinburgh had the only equivalent to a present-day agricultural extension service in Britain. The education it offered in adjacent fields, such as the department of 'ornamental agriculture', along with instruction in the philosophical tenets of associationism, provided lessons in aesthetic judgement. Principles of durable economy in farming were in contrast to the shifting tastes of gardening, the utility of form contrasting the conceit of style.[8]

Loudon left Scotland for London in 1803, in the first year of war with France, with letters of recommendation from his professor at the University of Edinburgh. He soon found influential clients.[9] His interest in the construction of forcing-houses is revealed in an 1805 treatise in which he describes several recent attempts to improve them.[10] His own experiments began with simple forcing-houses in his own backyard and later with prominent commissions, including the Palm House at Bicton Gardens.[11]

While describing the spherical forms best suited to capture the rays of the sun, George Mackenzie was first to admit that they were all but impossible to build.[12] Unlike the experimental equipment available to Stephen Hales and Jean-Etienne Guettard, the glass blower's skill was unable to produce a sphere large enough to suit Loudon's purposes – to say the least. Biographies highlight his invention in 1816 of a novel, wrought-iron sash bar that gave form to Mackenzie's idea (see Figure 3.4). Its profile was minimal compared to the timber used in traditional sashes, being only 5 cm (2 in) deep and 1.27 cm (½ in) in thickness and weighing only 454 gr (1 lb) for each 30.5 cm (12 in) of its length. These dimensions allowed the bar to be more easily bent into an arc. In his *Encyclopaedia of Gardening*, Loudon advertised the efficiency of his

3.3 Design for a new system for insulating forcing-houses, including a smoke flue for conducting heat through walls, in John Loudon, *A Short Treatise on Several Improvements, Recently Made in Hot-Houses* (1805)

3.4 Interior
view detailing
curvilinear
wrought-iron
sash bars used
in the Palm
House, Bicton
Gardens, Devon,
built by W. & D.
Bailey (1820–25)
according to
designs by John
Loudon

invention as a ratio of its thickness to its spanning capacity – one that exceeded the strength of nature's most 'economical' structure, the hen's egg – and its minimal obtrusiveness. Two years later Loudon transferred the manufacturing rights to his invention to W. & D. Bailey and began a series of collaborative projects with the manufacturing firm. Their last project together was the glass dome at Bretton Hall, Yorkshire in 1827. With a diameter of 30.5 m (100 ft) it was approximately half the size of Loudon's favoured scheme for the Birmingham Botanical Horticultural Garden four years later.

GUIDING GENIUS

Loudon's ideas on the form and appearance of iron-and-glass buildings as well as his manner of constructing them were informed by Scottish science. His thoughts on the need for serviceable shelter using rational structure and the questions they raised regarding aesthetic expression arose within a context where innovation was conditioned by social concerns. Concerns included recognition that ingenuity was itself a desirable character trait and needed to be rewarded. At the beginning of the nineteenth century, although novel production methods were becoming known throughout the iron-and-glass industries, there was still little common ground for research and coordination of information regarding new materials and their properties. The desire for wide, column-free spans and transparent surfaces, novel forms and the equipment to service, heat and ventilate them, required building components that were unlikely to have come 'off the shelf', as proved to be possible later in the century.

Inspired by the promise of a new class of plastic or malleable material, enthusiasm for iron was tempered in practice. Even before being wrought, cast and rolled, different types of ore and the kinds of products to be made raised different problems, each requiring considerable patience on the part of designers and great skill from the manufacturer.[13] Underscoring the novelty of both the material and the class of buildings they inspired is the fact that practices of invention, design and fabrication at the time were hardly clearly distinguishable compared to today. Working with these materials and buildings then involved a degree of risk and uncertainty that, today, is perhaps only associated with the highest of 'high-tech' industries.

The patent system arose in part as a response to such uncertainties. It transformed the grant of crown monopolies into a strategy for encouraging ingenuity with the promise of market protection and serving broader purposes of promoting British exports and industrial growth. Bentham saw the great utility of innovation and championed the system when he wrote: 'A man will not be at the expense and trouble of bringing to maturity invention unless he has a prospect of an adequate satisfaction, that is to say, at least of such a satisfaction as to his eyes appears an adequate one for such troubles and expense.'[4]

Despite its appeal on moral and commercial grounds, patenting was not universally welcomed. In fact, during subsequent debates on the reform of the system, Bentham came around to the view that the expense of patents proved to be one of the great burdens on 'men of industry'.[15] Regardless, patenting came to be an important aspect of the burgeoning exchange of materials and goods in nineteenth-century Britain. It served to underscore a context for new ways of making things, as well as to provide a medium for describing new things to be made, desired, purchased and consumed. It was also a means for relating the form and appearance of things to their underlying purposes.

It would be all too easy to attribute Loudon's failure to secure adequate patent protection for his innovations in wrought iron to 'misguided public service' or a lack of business acumen.[16] The significance of patents and intellectual property rights has meaning only within a framework of modern practices. Loudon's sash bar owed its notoriety more to the encyclopaedias in which it was advertised and the structures it formed than to any pre-existing competitive advantage. Likewise, the portrayal of the glasshouse itself as a unique, recognizable building type is largely a creation of hindsight. Loudon's works based on scientific principles coexisted with those premised on tradition, style and fantasy (see Figure 3.5). It was through such an ensemble of varied works that novel forms were distinguished from traditional ones and that expertise was set apart from the desires for the stylish and fantastic. Reliance on functional principles distinguished the experts who used them from those less disposed to do so; those architects, for instance, whom Loudon derided as having ideas that were 'by no means matured'.[17]

PROMOTING INGENUITY

The appeal of science to the 'mature' mind, the need to 'number, weigh and measure' and the language of functions and purposes, classes and types of objects were promoted through various means. These included scientific organizations and their popular counterparts in 'mutual improvement' societies, books and assorted kinds of publications. These media formed an important dimension of the social context for innovation in creating an audience ready to be informed of new devices, recent improvements and subjects both 'new and strange'. In 1826, shortly after his *Encyclopaedia of Gardening* appeared, Loudon's *Gardener's Magazine* became the first periodical for distributing knowledge of landscape gardening. It included articles culled from other works, such as his *Magazine of Natural History*, first published in 1828, and the massive *Hortus Britannicus* of 1830. The format of many of Loudon's works contributed to their success as publishing enterprises by allowing additional and updated material and experimental findings to be incorporated through several editions over a considerable length of time.

Just *how* to incorporate proven techniques and novel inventions, traditions and scientific advances required careful consideration. The Introduction to the *Encyclopaedia of Gardening*, for instance, refers to Samuel Coleridge's treatise on order that prefaced the poet's influential *Encyclopaedia Metropolitana*, published in the years 1817 to 1845. In this earlier work

Coleridge called for a 'way, or path, of transit' among the subjects of his work, as the only way to create a 'compendium of human knowledge'.[18] Working from the premise that the primary characteristic of reason was the superimposition of order upon the chaos of unrelated objects, the author of the *Metropolitana* proposed that order must come from the consideration of the relations between things, for:

> All things, in us, and about us, are a chaos, without Method: and so long as the mind is entirely passive, so long as there is an habitual submission of the understanding to mere events and images, as such, without any attempt to classify and arrange them, so long the chaos must continue. There may be transition, but there can never be progress; there may be sensation, but there cannot be thought: for the total absence of Method renders thinking impracticable; and we find that partial defects of Method proportionally render thinking a trouble and a fatigue.[19]

Coleridge's portrayal of reality as chaos differs markedly from that entailed in Locke's and Hume's respective accounts of human experience. For them, reason was generally believed to be a faculty that was distinct from sensory experience, though nonetheless reducible to the same mechanical operation of the mind. For Coleridge, however, reason and experience were neither independent of each other nor themselves reliable. The sensation of things brought only confusion, while reason was hardly as reliable as a well-oiled machine. Thought, he believed, did not arise spontaneously, but had to be trained or confined within the parameters of method. Relations between things, rather than things in themselves, were important, as the former were

3.5   View of the conservatory at Syon House, Middlesex, Charles Fowler, architect, 1827. New technologies were adapted to traditional building forms as well as those derived from scientific principles, as this 'architectural' conservatory illustrates

more amenable to observation, measurement and the discernment of causes and effects.

Loudon was equally concerned to establish a methodological rigour for his work as a landscape gardener, architectural critic and glasshouse designer. His methodical approach to built form was actively promoted along with inventions worthy of patenting, such as his wrought-iron sash bar, as well as other, less obvious inventions. These included his distinctive house for an invalid and even a new kind of garden cemetery. He too urged the consideration of relations between things, conceiving them as a kind of causal connection between forms and their purposes. These lay between the abstract measurement of sectional qualities of spaces or their manner of expression, on the one hand, and the beings who lived in them or derived pleasure from their forms, on the other. Loudon advised readers in that:

Works of art ... may be considered either in relation to their design or intention – to the nature of their construction for the intended purpose – or the nature of the end they are destined to serve; and their beauty accordingly will depend either upon the excellence or wisdom of the design, the fitness or propriety of the construction, or the utility of the end.[20]

THE LANGUAGE OF FUNCTION

The language of forms and functions, of purposes, fitness and utility that fills the *Encyclopaedia of Gardening* links the fascination with artificial climates in Loudon's day with a broader appreciation of the environment today. Invoking causes and effects conjoining buildings and landscapes to nature, this language allowed the reader to derive meaning from spaces. Interestingly this occurred in a manner akin to the use of microscopes, thermometers and other instruments by vegetalists in the eighteenth century or to the chemical nomenclature used for describing organic processes by scientists in the nineteenth. The language of function, in other words, allowed scientists and inventors to communicate new ideas and to describe how things worked. It allowed them to show those who were to use new ideas and objects the worth of scientific principles in solving everyday problems. In 1727, for instance, Stephen Hales approved of the contribution new instruments made to knowledge of plant processes:

It is now grown a common and very reasonable practice; to regulate the heat of stoves and green-houses by means of *Thermometers*, hung up in them. And for greater accuracy, many have the names of some of the principal exoticks, written upon their Thermometers, ever-against, the several degrees of heat, which are found by experience to be properest for them.[21]

In the domestic sphere the Victorian householder and gardener came equipped with a number of similar instruments which had, by then, become commonplace. Manuals of domestic economy began by stressing the need for thermometers, barometers and measuring devices of all sorts for assessing and controlling the unseen forces at work in the home. In being necessary for the operation of appliances for heating, ventilation and lighting and for controlling the supply of water – hot water to upper floors of multi-storey dwellings – these not only required consideration when planning the home, but constant care. As John Stevenson warned in *House Architecture* (1880), though the convenience of such labour-saving devices is great, 'the more there is of it, the greater the need for intelligent management, and the chance of its going wrong'.[22]

Chemistry and the use of quantitative methods, thermometers and the like related the specific tasks of the horticulturalist and homeowner to more abstract concerns for truth and objectivity, the efficient application of labour and the taming of uncertainty. The complex processes, organic and physical phenomena revealed by such methods in both curvilinear glasshouses and the Victorian home, cast horticulture and domestic economy as two of a number of enterprises amenable to science. Similarly, the language of forms and functions identified practices shared by biology, horticulture or landscape gardening, building or architecture. As a condition of clear, reasoned or methodical thinking, these terms defined each of these fields as a contemporary form of scientific understanding. This is not to say that fluency with such terms necessarily resulted in clear understanding any more than skill in quantitative methods was able to reconcile that 'ancient' antimony between phenomena and the 'noumenal' world – between things that can be sensed or perceived and those that are merely intuited. In fact, the discernment of causes and effects more probably forced these categories further apart.[23] In doing so, however, an important context for discrimination was opened.

By creating a distinction between landscape gardens and building forms derived from tradition or historic precedent and those based on scientific principles and methodical practices, Loudon affirmed Immanuel Kant's realization that human understanding was self-reflexive. In other words, we do not have direct access to gardens or buildings such as glasshouses as things in themselves, but only as objects of reflection.[24] We can reflect upon them using history, for example, or through scientific means. This notion of human self-reflexivity implies that reason does not assume one universal form; rather, that there may be 'styles' of reasoning and that objects are the product of reasoned practices.[25] The author of *An Encyclopaedia of Gardening*, Loudon, like James Fergusson in *A Handbook of Architecture*, may not have considered reason a style. In effect, though, by insisting on particular scientific or functional principles and the performance of certain mental operations as prerequisite to practice, they made thought dependent upon a distinct mode of experiencing the world. It was an experience where, above all, it was important to number, weigh and measure things, discern their purpose and discriminate between how they worked and what they looked like. It was an experience where built form was seen, as organic nature came to be, as infinitely malleable and subject to endless adaptation and adjustment to suit new contexts.

Loudon's encyclopaedias portray the history of design as the resolution of functional necessities by ever-resourceful humankind, thus making design, along with the need for places to shelter living beings – whether plants or humankind – 'facts' of history. The space Loudon sought to contain in his dome for the Birmingham Botanical Horticultural Gardens in 1831 and that Richard Turner and Joseph Paxton argued over in the Palm House in the decade that followed, drove their designers to ever-greater lengths of invention. Each spoke the language of functions and purposes, but gave form to the idea of the glasshouse in different ways. Each responded to the function or purpose of a glasshouse, though they failed to contain the environment entirely.

A necessary evil

With plans begun in 1843, the year of Loudon's death, the Palm House at Kew was completed in 1848, the year before Matthias Schleiden's *Principles of Scientific Botany, or Botany as an Inductive Science* appeared in English. The notoriety that weighs upon the building grew not merely from its celebration as a modernist icon in the twentieth century. It grew with the rise to prominence of the public botanical garden in the 1830s and 1840s, most notably Kew Gardens.[26] In broader terms and as the debates surrounding the design and construction of Kew's famous glasshouse reveal, the Palm House was – and still is – notable for the opportunity afforded by such a construction to situate human nature *and* culture within the broad realm of living species, animate and inanimate matter. These debates and the historical circumstances accompanying them are worth considering in some detail, not merely to show how, with hindsight, Loudon and his followers were mistaken or how the designers and builders of the Palm House got it wrong. Rather these events reveal how, in general terms, the built environment and its technologies create Latour's 'frame of reference' inside which biological events occur and human beings draw meaning from unique places.

By and large, histories of the Palm House celebrate its successes rather than its failures and the triumph of its designer with new forms and building methods. In some instances the circumstances attending its planning and construction have been cited to prove just *who* the designer was. In response to a longstanding controversy surrounding the authorship of the Palm House, Edward Diestelkamp researched the archives of Kew Gardens and composed an intriguing and detailed narrative entailing its design and construction and the principal figures involved. His account is worth elaborating here as it reveals how the desire to relate built form to nature by designing a functional structure to shelter living species was mediated by historical circumstance and the contingencies of practice.[27] What the Royal Commissioners and scientists associated with Kew Gardens and the architect and builder responsible for building the Palm House *knew* about nature and the function or purpose of glasshouses, in other words, was mediated by what they *did*, namely how they exercised their own forms of expertise.

It is worth considering the historian's narrative for another reason. Any expectation of finding a purpose or rationale guiding the evolution of the form of the Palm House there is thwarted. An assessment of the glasshouse's success as a truly functional building is mitigated by circumstances as unpredictable as the weather. Whether or not we can identify the hand of a single architect or engineer or a collaborative team upon it, we are no more certain why the Palm House did not function properly – at least not at first.

A REASONABLE APPROACH

The initial impetus for the overall scale and configuration of the structure followed a visit by the Director of Kew Gardens, Sir William Hooker, in June 1843 to the Duke of Devonshire's Great Conservatory at Chatsworth, completed by Joseph Paxton several years previously. Hooker asked the Clerk of Works to the Commissioner's of Woods for Richmond to prepare a design for a similar structure at Kew. Diestelkamp found a description of the brief in the *Gardener's Chronicle* of 30 September 1843: 'It is intended that this building shall be 200 ft. long (exclusive of approach or vestibule), 100 ft. wide and 55 ft. high. Like the Great Conservatory at Chatsworth, it will have a lofty centre surrounded by aisles forming (one interior), with a carriage drive through the middle.'[28]

A second scheme prepared by Hooker's curator, John Smith, had dimensions similar to those proposed by the Commissioner's clerk, with proportions, gallery and detail reflecting Paxton's model. Smith's concerns as a horticulturalist were consistent throughout the design and construction of the Palm House, being foremost the creation of a suitable enclosure for the care and protection of the plant collection. Little action was taken until the following January, when Richard Turner called on Hooker to discuss his own – the third – scheme for the new glasshouse. Obviously impressed, Hooker later wrote to his superiors that the Irish ironmonger knew 'more about hothouses and greenhouses and the best principles of heating them than any man I ever met with'.[29]

In February, in an interview with the Board of Commissioners, Turner formally submitted his plans as well as an estimate of costs for the work intended. It was then agreed that he should prepare more detailed drawings of the proposed range based upon this estimate. Attending the meeting by invitation of the Commissioners was Decimus Burton (1800–81), a collaborator in the design of the Great Conservatory at Chatsworth and architect to the Royal Botanic Society at Regent's Park, London. Like Turner, he also had 'acquired considerable experience in constructions of this nature'. The term 'architect' is elusive in reference to Chatsworth and is indicative of the nature of the controversy surrounding authorship of such novel buildings.[30] At this point it becomes interesting to note Burton's comments on Turner's plans, which reveal the differing aesthetic preoccupations of the professional designer and the manufacturer. Burton wrote:

The details of the internal arrangement and of the Plan appear to be prepared with much consideration, and to be well adapted to the practical object of the Building, but I cannot approve of the Elevations, the proportions nor the proposed mode of construction. The Ecclesiastical or Gothic style therefore in which the Elevations are designed appears to me to be objectionable, and the numerous ornamental details in fretwork, crockets, perforated parapets, etc. to be out of place and tending uselessly to increase the amount of the Estimate.[31]

While approving of the form of Turner's scheme, its provision of serviceable shelter for plants and the internal coherence of its structure, Burton found its manner of ornamentation out of order. Details of the scheme are scant and the form of its roof uncertain, though it may have resembled the palm houses Turner erected at the botanical gardens at Belfast according to a design by Charles Lanyon (1839–53), at Glasnevin, Dublin in 1842–49 and the Royal Botanic Society's Winter Garden at Regent's Park in 1845–46. These have sloping roofs, a form similar to those used by Turner's firm for the construction of gabled railway sheds. Additionally, they were fitted with operable sashes for controlling levels of humidity, a device also useful for ridding railway stations of locomotive smoke (see Figure 3.6).

Having horticulturalists as clients who were presumably better informed than himself on matters of science, the accommodating Irishman nonetheless incorporated into his scheme for Kew's Palm House curvilinear apses or rounded segments of roof where they joined walls.[32] Though it is uncertain to what extent Turner's proposal included Gothic features, Burton was clear in offering his own counter-proposal – the fourth so far – to suggest the limited value of such stylistic references. The form of his design was 'a conventional style suitable for horticultural purposes' in which 'the classic and ecclesiastical is avoided – appropriateness, with as pleasing an outline as the case will admit of, is the object aimed at, and all extraneous ornaments are dispensed with'.[33]

Correspondence suggests that Hooker favoured Burton's scheme at least in one respect, as Turner himself acknowledges:

as Mr Burton's drawing gentleman told me yesterday that you [have] given up the Roof sliding Sashes for ventilating and getting rid of the impure air from the summit of the house and being the exhalations of the plants I thought was indispensably important for the well growing of the various plants.[34]

Awareness of the possible danger posed by such 'exhalations' reminds one of the concern of the eighteenth-century vegetalist for plant respiration. In the same letter to Hooker, Turner attempted to dissuade him from accepting Burton's inclusion of a semi-circular roof for fear that the incidence of the sun's rays during winter upon the relatively flat top section would lessen the amount of light and radiant heat admitted. The comment supports a view of the mechanical behaviour of light likewise cultivated by vegetalists. The ironmonger's criticisms were furthered by John Smith, who attacked Burton's design on the basis of its structural incompetence, calling it an 'unmanageable affair'.[35] Smith was particularly critical of the absence of supporting lesser beams or purlin bars to carry the weight of the curvilinear aisle roofs, the interference of too many internal columns and its cumbersome trusses. He found the design 'to be wildly extravagant', claiming that 'its

interior will be much encumbered no doubt with a series of these immense massive trussed arched supporters'.[36]

In terms of the amount of free floor space it contained and its overall height, Burton's design was much more spacious than Turner's Glasnevin project, but as additional assurance against losing the commission, Turner wrote to Hooker that he was sending for his foreman from Dublin, 'a practical and Scientific' man who would solve the structural problems. Eventually, Turner, his foreman and Burton were able to propose yet another scheme – the fifth – which combined both a curvilinear profile and a clear span of 15 m (50 ft). It was a design which, allowing for minor developments, is reflected in the form of the Palm House today. The structural innovations which allowed this form to emerge must be credited to Turner and his staff. Significantly, the discourse of vegetality was invoked, though not by name, as the basis of its scientific character by all parties involved in the design process.

Turner substituted arched cast-iron ribs on approximately 3.5 m (12 ft) centres, supporting an infilling network of metal sashes and glass for Loudon's system of curvilinear sash bars which required additional supporting purlins and columns. This structure underwent several alterations as the design developed along with its construction, emphasizing the experimental nature of the technology. Immediately upon receiving news that he had been officially awarded the contract, Turner wrote to Hooker requesting the substitution of wrought-iron structural members for the cast-

3.6   Glasnevin Palm House, Dublin, Richard Turner and Charles Lanyon, 1842–49. Note the sliding sash openings in the roof for getting rid of 'impure air'

iron ones specified in the contract drawings. Turner perhaps realized the excessive costs of casting these ribs, each approximately 13 m (42 ft) long, the difficulty of erecting them and the danger of their collapse during construction. In an interesting instance of 'technological transfer' as it is popularly called today, the new wrought-iron structure capitalized on developments in shipbuilding where iron was quickly replacing timber as the material of choice and economy. Turner wrote to William Hooker on 23 November 1844, expressing his hopes not only for the revolutionary new structure, but for his own career:

I am getting all the requisites for satisfying *all* the Scientific Gentlemen, Mr. D. Burton and all of the vast improvement and utility and numberless advantages of the Wrought Iron *Ribs* and *Purlins* and our Roof framing. Which will be the most suitable of all and an *immense* advantage beyond anything almost conceivable in lieu of the tremendously prodigious and gigantic affairs proposed of which we shall have a most happy Riddance. I fully hope and *expect* with your efficient aid in Denouncing them not only as a *Blemish* but as a dreadful *Nuisance*, unsightly, unsuitable, and unnecessary and for Stability we shall be many *times* stronger than necessary *with our* beautifully light comparatively wrought Iron Framing, which I feel quite confident you will greatly *admire* and *approve* [emphases as recorded by Diestelkamp].[37]

As historians have argued, Burton's role was no less important to the design of the Palm House than Turner's. Reflecting the formal training and proclivities of an architect practising in the first half of the nineteenth century, his contribution is worth assessing not simply in terms of the extent of his influence. Rather, his role is interesting for the way in which 'scientific principles' served an aesthetic appreciation of form, composition and function. It is not that principles were applied inconsistently by either Turner or Burton in their designs, for there were no 'hard rules' for translating them into practice – for turning ideas about the function of a form into a novel form itself. This is obvious, not only given the debate concerning what the Palm House *should* look like and the relevance of style and ornament in a building serving horticultural purposes, but in terms of an apparent disagreement regarding the positioning of the building itself.

PROBLEMS OF FORM

The original site for the glasshouse determined by Hooker in February 1844 was on the north side of the pond, near what is now the Garden's Museum at the end and on the west side of the broad walk that ran past the orangerie (see Plate III). Hooker's motives for such an orientation are not clear. For such a long building, orienting its main axis perpendicular, rather than parallel, to the path of the sun would seem to contradict Mackenzie's advice concerning the orientation of earlier curvilinear or nearly curvilinear forcing-houses. Most notably, the orientation counters that of the Palm House at Bicton Gardens and Turner's Glasnevin range. Unlike these structures, however, designs for the Palm House at Kew were glazed on all sides. They lacked the masonry wall that would have formed a northern 'back' to the forcing-house and served to absorb the sun's heat. Given improvements in heating systems, the provision of this wall was no longer necessary, thus affording greater flexibility in the positioning of glasshouses.

The placement of the Palm House at Kew appears to respond to the prominence of gardens as a public institution and the desire to present its buildings and collections from a suitable vantage point. It appears to respond more to aesthetic than scientific purposes. In April of 1844, Burton staked out the site along the west side of the pond so that viewers might 'gain the reflection of the building in the water'.[38] Apparently this proposition was accepted without due consideration of an alternative arrangement proposed by John Smith two days later. Smith later blamed Burton for the frequent flooding of the furnace area beneath the floor of the Palm House, due to the position of the building next to a pond.

Regardless of *who* designed the Palm House in Kew Gardens, the reasons *why* its botanical arrangements suffered despite the involvement of so many 'practical and scientific' men is an interesting question. While flooding was not responsible for killing the palms in the Palm House in 1867, the reason for their pallid fronds is no more certain. One should recall that the orientation of Mackenzie's forcing-house relative to the fixed position of plants within and the east–west axis of the sun's transit, along with an awareness of the obtrusiveness of structure, were key concerns of the horticulturalist with an up-to-date knowledge of vegetable life. It had been commonly observed, however, that plants close to the glass panes of ordinary houses were often scorched. A number of explanations for this phenomenon were given, entailing, in some instances, imperfections in the glass, but in others, irregularities in the force of sunlight. Loudon had observed in 1822 what 'every gardener knows', that plants would not thrive if they were placed some distance away from the surface of the forcing-house. He acknowledged James Sowerby's work on light and colour in 1816, speculating that the type of glass commonly used in greenhouses often behaved like a prism to 'derange the order of the rays'.[39] Thirteen years later, in the fifth edition of his *Encyclopaedia of Gardening*, Loudon alerted the reader to the potentially harmful effects of the green-tinted crown glass:

Economy as to the quality of glass, therefore, is defeating the intention of building hothouses which is to imitate a natural climate in all the qualities of light, heat, air, water, earth etc. as perfectly as possible. Without a free influx of light, the sickly pale etiolated appearance offends any who take an interest in the vegetable kingdom.[40]

If the transparency of glass and its effect on plant life proved vexing, further problems shadowed the curvilinear forcing-house. Whereas Mackenzie's innovation was to conceive of the glasshouse as an immense convex lens capable of minimizing the reflection of light away from its interior, others considered the 'behaviour' of light as it passed through the glass and was refracted. They observed how the concavity of the structure's surface acted like a magnifying glass to burn tender leaves and concluded that sloping roofs were preferable to curvilinear ones as the concentration of light was avoided. As a device by which the sun's force was viewed simply, narrowed into rays of visible light or heat and accordingly measured, the curvilinear forcing-houses fabricated by Turner and his chief competitors in the glasshouse trade conceivably could have induced an excess of light as much as their glass might have prevented sunlight from entering. Whereas Sowerby's research suggested to Loudon that chromaticity was potentially a problem to be overcome as regards the convergence of light rays, their 'purposeful' derangement might actually prove useful. At the request of William Hooker, and the advice of John Lindley, the first Professor of Botany at London's University College, for instance, the great curved, iron ribs of the Palm House were fitted with green-tinted glass as a means of preventing plant scorching, or so it was hoped at the time.[41]

Further uncertainty and vexation accompanied the vegetalist's belief that nature could be made to perform like clockwork in the conservatory. Without a clear understanding of cellular metabolism that could explain the sun's role in supplying nutrients and assisting respiration in common terms, some experts believed that plants gave off quantities of 'phlogistonated' air that was harmful to both plants and humans if not removed from enclosed spaces. Turner's efforts to remove the threat of 'impure air' from his glasshouses further emphasized the mechanical nature of the sun's force – here, in terms of the process of respiration – as a direct relationship of cause (the light of day) and effect (the emission of phlogiston). More significant, however, his concerns involved an implicit acknowledgement of the multiple reactions a singular force might induce: the beneficial warmth of the sun's rays and the impure air which plants emit upon exposure to sunlight. Each effect compelled an entirely different technological response: a curved roof to capture the sun's rays versus sloping roofs with operable sashes to prevent the burning of leaves and to expel impure air. The climate of doubt that accompanied speculation of the multiple causes of natural phenomena and acknowledgement of the possible effects of various technological means for remedying them made the artifice of artificial climates truly challenging. Three decades earlier, in fact, Mackenzie had expressed uncertainty as to whether the glass in glasshouses exerted a positive or negative influence on the species sheltering beneath it, a question which was to be settled by yet further investigation.[42]

AN 'UNMANAGEABLE AFFAIR'

Joseph Dalton Hooker inherited this climate of uncertainty along with his father's job as Director of Kew Gardens. This was largely because of the lingering influence of an outmoded form of reasoning and an untested building form that could not be tested for it was largely an experiment in accommodating nature. Records show that the Palm House was indeed reglazed with clear glass. For John Hix, this was a response to the fact that the 'plants did not flourish for lack of light'.[43] Joseph Hooker's letter to his superiors, describing the 'radiation of cold from the iron and glass' was a response to a report on the structure's deficient heating system, which he received in August 1867. His answer was to introduce two flue chimneys in the wings of the house. Punctuating its now famous curvilinear profile, this was one of several improvements to the heating system that have continued to the present day.[44]

Requiring no less attentiveness than heating and proving a 'dreadful' nuisance equal to the timber structures they were designed to replace, Turner's ensemble of iron ribs and purlins enveloped the horticulturalist with yet further disaffection in both plant and human terms. It proved to contain not noxious phlogiston, but rather high levels of humidity. The dire consequences of such conditions for plant life had been known for some time and were a problem common to forcing-houses, not just the scientifically designed ones, as Williams had previously observed.[45] Without proper ventilation, humidity was not only thought detrimental to plants, but proved harmful to the structure itself, causing it to deteriorate and thereby necessitating expensive repair. Even before the rust began to bloom amidst the luxuriant forms of the 'Mango, Cocoa-nut, Chocolate and Breadfruit' in the Palm House, their long-suffering caretaker, the curator and failed hothouse designer John Smith, found a scene as objectionable as the vulgar pagoda Louis Le Comte discovered in eighteenth-century China:

In looking up nothing but iron was to be seen in every direction in the form of massive iron, rafters, girders, galleries, pillars and staircase, and the hot iron floor on which we stood and the smooth stone shelves and paths round the house had more the appearance of some dock-yard smithy or iron railway station than a hothouse to grow tropical plants in, but there it was, and I was to make the best of it, and to be responsible for the good cultivation of the plants which were commenced to be put in.[46]

Given the apocalyptic scene to appear in the Palm House almost twenty years later and in the face of its uncertain cause, one might sympathize with Smith and agree with Mary Woods and Arete Warren as they looked into the past and into the glasshouse with disbelief, observing how it was as though the palms themselves 'refused to grow'.[47] Perhaps the plants, like their curator, found the form of the modern glasshouse an entirely 'unmanageable affair'.

If the mechanical view of nature inherited from eighteenth-century science proved unworkable in the Palm House, its fate was sealed in the adjacent Temperate House, construction of which was initiated in 1859. This building is not only an instance of what might commonly be understood as applied science, but also an economy of ways and means, of the needs of plants but also of society. As such, economy was not necessarily limited to the arrangement of the building's plan, its construction and maintenance costs, but incorporated aesthetic directives as well (see Figure 3.7). The building brief specified that it

be not curvilinear so that large operable sashes on a straight roof could be employed to allow maximum exposure to summer rains and temperature. That these sashes be made of wood as they could be more easily repaired by garden staff than iron ones could, and that the bald edge of the roof be hidden by a stone cornice.[48]

3.7 The
Temperate
House, Royal
Botanic
Gardens, Kew,
by Decimus
Burton,
construction
initiated in 1859
with subsequent
building
additions

The conflict, between the desire to use scientific principles to create functional forms and the consequences of new technologies that did not always work as they should have, characterizes the period encompassing the building of the Palm House. This is a conflict reflected in our experience of the built environment today, though we no longer believe in the discourse of vegetality but rather in biological science. Beginning in 1843, the use and implementation of scientific principles and the submission of expert designs by various individuals provided important media for experimentation and debate. This facilitated the transformation of more general attitudes to nature. In the Palm House and in other forcing-houses, observations of the effects of coloured light on plants led to thoughts on photosynthesis and its definition as a process involving chemical and physical reactions at a cellular level. The attempt to explain colour in terms of light of varying wavelengths led to the demise of corpuscular optics and the growing influence of physics and thoughts on molecular excitation. Similarly, Priestley and Lavoisier initiated a series of investigations whereby new methods of chemical analysis ultimately displaced phlogistic theories of gas. These scientific advances allowed analysis and understanding of the products of plant respiration and the analysis of soil relative to nutritive processes to be quantifiably determined.

A consequence of such practices was that the forces of nature studied by Classical scientists were multiplied. A complex relationship between an organism and its environment subsumed mechanism. It provided the basis for continued technological intervention in the natural world. Causes and effects, both positive and negative, were subsumed by unitary theories of the cell and of matter as science now sought to relate decay and vitality, illness and health. These came to be tied to one another via reference to a third set of variables, including the adaptation of organisms to their immediate environs. Scientific practices of observation, the hypothetical modelling of phenomena and forming of deduction whereby Loudon charted the amount of visible light passing through various glazing systems were applied to an entire range of factors governing plant vitality. This was done in an attempt to address technologically the balance presumed to exist between plants and the environment. The brief for a particularly 'modern' structure such as the Temperate House allowed for a coherence of form and space by defining its purpose – or purposes relating both to nature and to social needs – while nonetheless remaining technologically flexible.

A climate for invention

While following the shift in attitudes towards nature from one period to another, we should not assume too direct a correspondence between changing theories and technological practices. As the history of the curvilinear glasshouse shows, relations between these two domains of human enterprise were mediated by historical circumstance. History can account, for instance, for why generally the study of vegetality took a back seat to agricultural practice during the seventeenth and eighteenth centuries. Studies of plant physiology proceeded more swiftly in the early years of the nineteenth century in France and Germany than in Great Britain. This is despite the fact that state involvement in agricultural production began at an earlier date under the Georgian monarchy. A strong, central government allied to business interests in Britain prolonged interest in systemic or taxonomic studies and fostered the discovery of new plant species throughout the empire. Interests in more abstract studies of plant perspiration, respiration and nutrition took a back seat. In Germany, on the other hand, numerous principalities and their universities were responsible for furthering research in speculative, botanical science during a period of political and economic rejuvenation following the defeat of Prussia by Napoleon in 1806. This had the effect of exposing Germany to greater French influence in science and industry.[49] Schleiden complained, for instance, that the study of complex plant species and botanical systems hindered the development of physiology, which was more evident in simple, individual specimens.[50] The British interest in the former type of study and the advanced state of British industry begins to explain why the Palm House at Kew, by the time it was built, embodied fairly outmoded ideas about vegetable life although benefiting from advanced construction practices.

Historical circumstance can further explain why, at a given time, it was more likely that innovation occurred in some places and not others. As development of its industry had been hindered by crown monopoly, the newly independent United States of America would have been likely to remain dependent on foreign sources of iron and glass. One would not expect to find curvilinear forcing-houses in that country although certainly Loudon's principles would have been known there. It was more likely for iron-and-glass structures to be larger, earlier, in Ireland than in Scotland and England during the 1830s as Ireland was exempt from certain taxes on glass as part of British export schemes. A consideration of varied historical circumstances suggests that the fortunes of empire, governments, science and industry are neither simply determined nor in themselves completely *determining* of ideas and practices.

The energy expended creating the Royal Botanic Gardens at Kew and its Palm House spread across the entire metropolis in its public parks and squares to ameliorate the worst excesses of the industrial revolution.[51] In his survey of London's *Parks and Pleasure Grounds* (1852), Charles Smith discovered

all classes of the community, the day-tasked official, the night-worn senator, the slaves of business, and the votaries of fashions, even royalty itself, all availing themselves of the air and exercise, and scenes of gaiety and opportunities of social intercourse and enjoyment which these much frequented places afford. Nor is it to be overlooked that the public parks, and even the smaller gardens in squares and streets, are fitted, if skilfully distributed, to lessen the condensation of our large cities, to extend their crowding buildings over a wider surface, to rarefy the thick black clouds of smoke which rise from them, and so to increase their light, and to provide a lager supply of salubrious air for all the inhabitants. In short, they are, as it were, the lungs of cities and towns; and as such they are breathing-places to thousands who may never wander from the streets within their actual precincts.[52]

Like Joseph Paxton's Crystal Palace, completed just one year before Smith's survey, London's network of parks provided fertile ground for thinking about society, built form and natural environments. Above it all, though, was yet another factor that overshadowed the cultivation of reason in the glasshouse, London's parks and pleasure-grounds and the house and garden. A source of endless commentary and constant anxiety, it was a concern that, wherever they plied their skills, Britain's conservatory designers, architects and horticultural experts always took with them. It was the English weather.

Conceivably, the peculiarities of Great Britain's geography and meteorology enhanced the allure of artificial climates. Aggravating the appalling, sulphur-laden fogs of the nineteenth century, perhaps the fading sun at the heart of empire was responsible for killing the palms the Palm House was meant to nurture. The 1830s and 1840s was not only a period of intense construction in iron and glass as heralded by the great public conservatories established at Dublin, Glasgow and Kew, but was one that saw Europe's 'little ice-age' exhaust itself with a final exhalation of murky, foul-smelling air.

Kew Gardens not only set the scene for death in its most famous conservatory in 1867. It had also been, for some time, one of two key positions for recording statistics relating to climate for Greater London. The other was the Royal Observatory at Greenwich. From the time of Roman occupation to the early Victorian era, evidence for changing weather patterns in London was mainly descriptive. Temperatures were occasionally recorded from the beginning of the eighteenth century onwards. Rainfall records are available from the seventeenth century, though an account of total hours of sunshine for a given month commenced only in 1881. Fog reports are difficult to interpret since remarks on visibility were not standardized until 1920 and observations throughout the 24 hours were not made until 1941.[53]

Searching the record for illuminating coincidences – or rather, dark episodes – in the story of London's weather, it is worth noting that the worst frost of the nineteenth century commenced on the evening of 27 December 1813. It was followed by a dense and famous fog that persisted for seven days, bringing the city to a standstill. Had the fog persisted even longer, alternative routes for travel across town would have opened when the Thames froze completely solid on 31 January 1814. The cold spell persisted and was followed by another particularly severe winter. This was itself followed by twelve months 'without a summer' and the year in which Mackenzie delivered his essay on greenhouses. This was shortly before Loudon's call for artificial climates. Admittedly, there is little to suggest that the winter immediately preceding Joseph Hooker's comments in 1867 on the tragic state of his palms and vines was a particularly severe one in terms of mean temperature. There were other aspects of weather, though, that should be considered.

If it was not necessarily the absence of the sun's warmth that failed to 'cook' Kew's prized botanical collection, perhaps it was something else serving to 'derange the order' of the sun's rays. John Brazell notes that for much of the city's early and medieval history, Londoners, like people everywhere generally, did not travel out at night and when they did they proceeded slowly. Fogs were not so much a problem and commented on less frequently than, say, in the fifteenth and sixteenth centuries, a period of increased population and atmospheric pollution produced by the burning of coal.[54] The burning of coal increased markedly, of course, during the nineteenth century and alterations to the Palm House included two additional chimneys to support the thickening sky over Britain. Although London more or less warmed up for the Victorian period, its fogs got much worse (see Figure 3.8).

The meaning of fog was not standardized until after the First World War, when the development of aviation made its assessment a matter of death, not only lifestyle – it being essential to see then in order to fly.[55] According to available accounts, whereas Greenwich experienced on average nearly 19 days of fog per year in the second decade of the nineteenth century, this increased to nearly 55 days per year for the period 1881 to 1890. The foggiest winter, according to one source, was 1890–91, when fog was reported on 50 days. The *Kew Bulletin* for February 1895 reported that by the end of the century the increasing haziness of the sky caused by the rapid extension of London to the south-west made the use of tinted glass no longer necessary. The 'extreme obscurity' of the winter of 1885–86 in particular proved that 'no available sunlight could possibly be spared'.[56] According to one estimate detailed in the *Bulletin*, green glass used at Kew intercepted about half the influence of ordinary sunlight on plant life, according to the 'modern accepted data of vegetable physiology' at the time.

If events in 1867 cast a pall over Loudon's hopes for artificial climates, then William Dallimore's *Reminiscences* were no less inspiring. Employed as a gardener at Kew Gardens from 1891 to 1936, he arrived there during the particularly unpleasant London winter and was treated to a view of the Garden's conservatories through glass stained by 'greasy soot and other filth'. It was a torrid zone of a different kind, one far removed from the exotic 'Hindostan' of Gosse's wanderings:

I entered the gardens by the Main Entrance. It was a dull, damp, messy kind of day, and everything looked dirty and dismal. The Aroid House – the first seen – was not inspiring; it looked as though it was covered with slates instead of glass, but it was not until I saw the Ferneries, Greenhouse, Succulent House and T-Range that I fully appreciated the dirt. The glass had not been washed after the fog and it was black with filth. To make things worse the ferneries were wholly, and some of the other houses partly glazed with green glass, and most of them needed rebuilding. I did not know then what a harmful effect London fog had on plant life, and was not prepared for the many leafless and flowerless plants that should have been in first rate condition. Neither did I know of the scheme of reorganisation that was to transform the whole face of the establishment within the next few ears. As I made my way to the Temperate House the filthy state of the leaves of evergreens was noticeable on all sides, while small pools of water collected on the paths were covered by a black film.[57]

As the history of the Palm House reveals in the years after its completion, there was no room for a simple view of nature. The environment was a far

A LONDON FOG.—DRAWN BY DUNCAN.

more complex entity than nature could ever have been in earlier times. At Kew, there could have been no more dramatic a barometer of the extent of environmental forces that raged around it than its famous iron-and-glass structures. On 3 August 1879 gardeners recorded the first instance of hail damage there, observing stones so large they struck with a force that buried them in the ground. Having had a small percentage of the Gardens' total extent covered in glass by the time of the storm, its magnitude could be measured in a most spectacular fashion, by 38,648 broken panes or 18 tonnes of shattered glass.[58]

During the first half of the nineteenth century, the common forcing-house was transformed from an object of horticultural interest into a unique building type. The glasshouse became an analogue for an infinitely complex, it was hoped malleable, if potentially malevolent, world. Accompanying its rise in popularity was an expanded field of needs, opportunities and liabilities. It was a context formed from agricultural crisis and the concerns of manufacturers and politicians, economists and social reformers. It incorporated new modes of producing and consuming natural resources and new ways of making things. It was a context in which human identity was multi-faceted, in being both bound to nature and responsible for transforming the natural world at the same time. The glasshouse and the

3.8 'A London Fog – Drawn by Duncan', *The Illustrated London News* (January–June 1847)

environment it was designed for, though never fully contained, were a means for reflecting on both prospects.

Supported by government policies and patents, scientific practices and the books that promoted them, the imperative to *think* about the environment in a methodical way, to imitate or improve it, was an important catalyst for social change. It drew the reformer and engineer, the architect and landscape gardener to reconsider horticultural buildings like forcing-houses. It also provoked thoughts on easily overlooked places such as mines, gaols, hospitals, workhouses and ships – even cemeteries. While the improvement of these places served various, commonly espoused philanthropic ideals, the outcome was the same. It was their incorporation into an abstract domain of potentially habitable, useful and always thought-provoking space.

The appeal to science not only helped residents of Britain's cities adapt to their surroundings. It led them to become more valued, productive and creative members of society as well. To understand the function or purpose of a thing, whether living form or human-made object, was to encourage individuals to situate *themselves* within animate and inanimate nature, to care for and use it. Understanding the purpose of things was not a given – the gift of universal reason. The workings of organic and inorganic matter required certain acuities and powers of discernment to appreciate them. The growth of distinct forms of expertise and the technologies they inspired accompanied industrial growth in many European countries in the nineteenth century, but was particularly evident in Britain. It was then expressed through a commonplace view of Victorian society as complex and structured by its professional and learned classes, comprising horticulturalists, architects and engineers among others.[59] In 1857 Henry Thomson advised in *The Choice of a Profession*: 'The importance of the professions and the professional classes can hardly be overrated, they form the head of the great English middle class, maintain its tone of independence, keep up to the mark its standard of morality, and direct its intelligence.'[60]

The turn to science in the glasshouse prefigured its use in the home. Awareness of the workings of organic and inorganic matter, the imperatives to improve one's environs or adapt oneself to them, broadened the search for the order of things, their causes and effects. It came to encompass the house and garden, but then also the world outside the home: its geographical situation, the earth on which it stood, the air its occupants breathed and the water they drank – even its weather.

Elemental existence

The 'habitat' housed by Loudon's designs for the Birmingham Botanical Horticultural Garden, though speculative or intuitive in 1831, prefigures our understanding of the term today. Loudon's proposal was to place plant and animal species within a single building form – in a unique place. Though perhaps not yet representative of specific geographic localities or 'real' climates per se, but rather emblematic of geographically determined space, his scheme prefigures the purpose of a modern botanical garden. The idea that humans might enter such a domain and by observing the life forms within learn something about themselves transformed the garden of the apothecary or natural historian into a public museum – a means for moral and social improvement. An appreciation of the pace of scientific and technological development to occur over the course of the next several decades may be had by noting the appearance of another term associated with environmental science: 'ecology'.[1] In 1866 the term was used for the first time to describe a broad field of study concerned with the interrelationships of living organisms and their surroundings.

This field was formed by terms such as habitat and ecology and by new attitudes to nature that are now commonplace.[2] Awareness of the dependence of living beings upon their surroundings – particularly human beings upon their urban environs – was equally cultivated in novel architectural spaces and landscapes in nineteenth-century Britain. Botanical gardens were but one example. This chapter describes environmental awareness as a consequence of life in such spaces as they encouraged new ways of thinking about one's surroundings, particularly nature's fundamental elements of earth, air and water.

In Britain opportunities for observing the influence of nature on urban residents were varied and included spaces large and small, public and private. At one extreme was Joseph Paxton's Crystal Palace of 1851 and his proposal four years later for a 'Great Victorian Way', a vast arcade of shops, apartments and thoroughfares that was to encircle central London (see Plate IV). Lynda Nead describes how the project was proposed to ameliorate the vicissitudes of the city's climate, smoke and dirt and to solve its traffic problems. Its major social impact, however, was to further the control and segregation of public space.³ One of several, similarly scaled proposals, the street of glass capitalized on new modes of perception and discernment and new forms of urban experience.⁴ Paxton's unrealized design was complemented by a network of less grandiose, though built, works including glasshouses such as the Palm House at Kew Gardens, shopping arcades, railway stations, and exhibition and public halls. In each the appreciation of unique places was furthered as their occupants were led to consider the scale, enclosure and ambient qualities of spaces and the magnification or exclusion of the city's weather, pollution or noise. These structures made residents aware, not only of the qualities of other, urban or seemingly more 'natural' surroundings, but of the potential for human beings to alter or improve both.

At an even lesser scale – though forming a substantial portion of the fabric of British cities by the second half of the nineteenth century – the house and private garden were enclosures in which the forms and forces of the natural world were made known, isolated and measured. The very novelty of environmental awareness is a key aspect of urban experience, being one means by which the activities of life became organized in the domestic realm. In this regard the vital landscape enclosed by the botanical garden and glasshouse can be extended to include the Victorian home. It was in the home that nature's beneficial or harmful influences were enhanced or remedied – having been first observed as so many causes and effects. These influences bound the residents to their immediate environs and to broader domains, to neighbourhoods, cities and the nation as a whole. Being aware of the environment, however, was not always a good thing. The unknown origins or consequences of dampness, decay or filth provoked considerable anxiety for residents. Fears of disease and even moral ruin followed. The desire to forestall such eventualities encouraged the fortification of city and home alike against pestilence, impurities and poisons that threatened the vitality of life itself.

Attentiveness to the environment was given a boost with the publication of landmark texts such as Thomas Malthus's *Essay on the Principle of Population* (1798), Charles Darwin's *On the Origin of Species* (1859) or George Perkins Marsh's *Man and Nature* (1864). However, these do not fully account for the sense of awareness intended here. These texts, and others exhibiting a particular regard for geography and the distinctiveness of specific locales based on the life forms found there, provided a philosophical gloss to widespread beliefs in natural order and the interdependence of species and animate and inanimate matter. Less celebrated works that often claimed a basis in science, while professing to be of 'practical' value, introduced the householder to nature's fundamental elements and operations close at hand. These included diverse literary genres, comprising books and journals of popular science, treatises of domestic architecture and books on gardening as well as manuals of household economy – even fiction. They were means by which sensitivities were directed to issues of everyday life and the places in which contact with nature occurred on a regular basis.

The tone of this chapter is best set by a book of popular science and household advice of the Victorian period. Cuthbert Johnson was a prolific commentator on agriculture and home economics. His book *Our House and Garden: What We See, and What We Do Not See In Them* (1864) brought the advances of science to the attention of the average British householder. Not only was environmental awareness explicitly celebrated, if not named; it was turned to imaginative as well as practical purposes. The author described, among other phenomena, the reason for feeling cold upon arising from bed, the dampness of different materials from which linen and bedclothes were made, the porosity of walls and water-tightness of masonry; the composition of the air, its movement, freshness or closeness, and the smells of objects within confined spaces. Upon leaving the house and entering the garden, a 'new class of phenomena, other abounding marvels' greeted the viewer:

Of these, many are apparent to our senses – we might remark them at the first glance, if we but *think* a little – many other wonders belong to the unseen movements of the vegetable world. We have recourse, therefore, to the chemical philosopher to explain to us in some degree hidden mysteries that we can otherwise only know by the results they produce.[5]

In Johnson's work the house and garden were thought of as a contiguous whole – an environment. Readers were invited to learn something of biology and the natural sciences, physics, chemistry and even meteorology, as these fields revealed the myriad of organic and inorganic operations occurring in the home. These fields bound interior and exterior spaces into a common domain where nature caused things to happen and where natural forces like gravity, chemical reactions and the weather all had their effects. By appreciating the reasons for bodily sensations and potential sources of discomfort, the householder acquired an ability to interpret their surroundings. While Johnson illustrated the many practical applications of science, his text demanded that readers exercise imaginary powers equal to the seemingly unlimited opportunities posed by the use of science. Books like *Our House and Garden* called upon science and the unique characteristics of household spaces and garden landscapes to reveal – to make real – domestic environs that were the source of novelty and surprise.

It would be naïve, of course, to assume that works of home economics, domestic architecture and gardening served mainly to cultivate curiosity. It was obvious by the century's end that they also served to define domestic labour. They came to describe tasks suitable for the female homemaker and the male householder. More generally, they encouraged respect for enterprise along with inquisitiveness. In language typical of domestic science, one period textbook describes the homemaker as a skilled worker who could ill 'afford to be ignorant of science, since it is only by conforming to the laws of science that she may hope to do her best, whether in cooking, in washing, in ventilating, or in disinfecting'.[6] The dutiful regard for these activities not only reflected the character of the home's occupants as each was assigned particular tasks according to gender, age and class. These activities also cast the family residence as an exemplary vehicle for scientific experiment and self-reflection. Cartwright's textbook of domestic science of 1900 based a thorough knowledge of household economy on such basic practices as weighing and measuring. It exposed the homemaker to ideas regarding the density or specific gravity of materials and the temperature and movement of air. It taught them that they too were affected by the circulation, solubility and chemical composition of liquids among other physical phenomena and organic processes. It was in similar terms that Stephen Hales had said of the activities in the laboratory, in the pursuit of both pleasurable activities and demanding labours, that 'the most likely way ... to get insight into the nature of those parts of creation, which come within our observation, must in all reason be to number, weight and measure' (see Note 4 in Chapter 2). By thinking about, numbering, weighing and measuring nature's elements and forces, the home became a place where nature *might* be mastered, but only with considerable effort.

Departments of nature

Developments in science, particularly biology, gave philosophers and social reformers reason to believe that living beings were influenced by where and how they lived. For those interested in how plants and animals 'worked', science suggested that the study of organic structures was not merely of interest for its own sake – for presenting to the eye a number of exotic forms. Rather, the study of physiology was valuable for revealing how organisms were adapted to suit their immediate environs. Accordingly, the environment appeared to comprise a myriad of 'causal' relationships formed between living beings and their surroundings. This invited thoughts on the function of a given organ or the cause of a particular sensation, the operation of physical nature or the behaviour of inorganic elements and compounds. These issues were no less important for humans than for other living species. Given the growth of European cities, particularly in Britain by the middle decades of the nineteenth century, the impact of environmental factors on the body was no less evident or talked about in the city and home than it was in the field or laboratory.

In *Our House and Garden*, for instance, Johnson described the freshness of air on the skin as a function of ventilation. This was further dependent on windows of a certain size, while a sense of dampness was a consequence of particularly porous building materials or soils. This manner of reasoning – of distinguishing causes from effects – pervades the entire book. While the factors that produced sensations like freshness or dampness may have seemed invisible and mysterious, the householder and gardener were able to appreciate them by observing both the impact they had on their own bodies *and* the surroundings within which these phenomena arose. It followed that they were able – in theory at least – to alter the course of events themselves, encouraging the one sensation and preventing another less desirable one. The anonymous author of *The House and its Surroundings* (1878) admitted that 'the business of life, whatever in each case it may be, compels most of us to live in a particular locality, and to take the soil, air, water and houses of that locality as we find them'.[7] This was not to suggest that residents need be subservient to the demands of nature. Rather it was a call for diligence and enterprise, for everyone 'in an individual as well as a corporate capacity, can, in the administration of his own house and its surroundings, do much to influence his own health as well as that of his neighbours'.

For the Victorian householder, like the scientist, earth, air and water provided obvious points of communality between human beings and lesser species, plants and animals based on processes of nutrition, respiration and reproduction exhibited, in some form, by all living things.[8] In terms of everyday experience, attentiveness to earth, air and water in the house and garden served not only to reveal these vital processes; it also made evident the patterns of life shared by them. In this regard garden plants and domesticated animals and even living 'poisons' like germs were part of organic systems in which they interacted with one another. Anxieties regarding elements and organisms in the domestic sphere, both good and bad, provoked the occupants of Britain's residential districts to situate themselves within the broader landscape of organic existence. They were obliged to organize their lives within environs ranging in scale from the local to the global. A scrutiny of soil and dust, air and water close at hand was complemented by more broadly scaled studies, whereby nature was cast in geologic, atmospheric or hydrological terms.

At the scale of residential districts, for instance, John Loudon's *Suburban Gardener and Villa Companion* of 1838 came with a plan for determining the desirability of homes relative to their position alongside streets as they were commonly set out in towns and suburbs (see Figure 4.1). It was a diagram for representing the 'aspect' or orientation of houses and front, side and rear gardens relative to the sun's passage from east to west. It was reminiscent perhaps of Mackenzie's concerns years earlier for the orientation of glasshouses. In Loudon's diagram, however, there arose multiple concerns stemming from the exposure of buildings and gardens to sunlight *and* shade. Readers learned that lawns were likely to remain damp and walkways dry, given the relative exposure of each to the sun. The prospective home buyer was led to consider the species best planted in some places and not others. If neighbourhoods were surrounded by dusty roads or a smoky atmosphere, however, they were told that the careful choice of a building lot would matter little.[9] The diagram was a useful tool for Victorian homebuyers, helping them imagine the many promises and pitfalls of new abodes. In doing so, it represented their houses and backyards, surrounding neighbourhoods and beyond as a contiguous whole. Loudon may not have had an environmentalist's viewpoint per se, though he drew on his experience of sound building and horticultural practices in a methodical way to invoke a domain of physical and organic operations to which residents must attend. The many details of this domain were added by scientists in the ensuing decades. Loudon's intimations that the home was not a closed world, but one

4.1 Diagram representing the four leading directions of streets occurring in towns and suburbs and the positions of houses and gardens most favourable to the admission of the sun throughout the year. Note that shadows are not cast at one particular time, but represent areas where little or no sunlight will fall throughout the year. In John Claudius Loudon, *Suburban Gardener and Villa Companion* (1838)

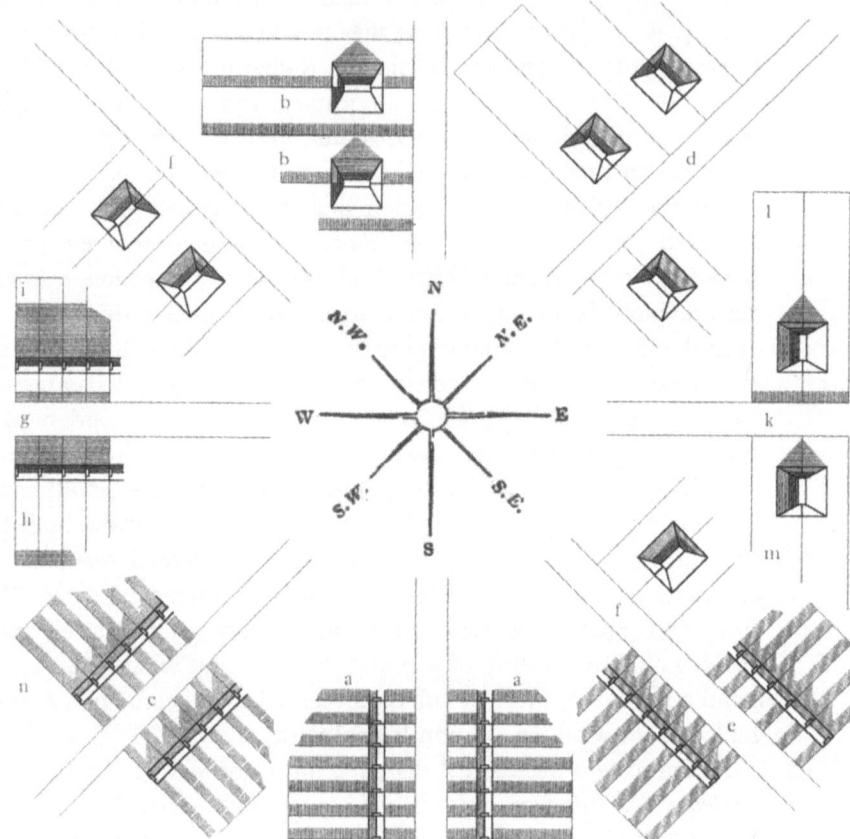

that opened ever outward, was supported in 1849 by John Davy, who extended his study of chemistry to encompass the 'three great departments of nature' into which he divided the habitable world: namely, the atmosphere, the earth and the ocean. He was particularly concerned to question the way in which these were connected and 'administer to each other in a marvellous manner'.[10]

In this section earth, air and water are isolated in turn to survey some of the ways in which they impinged on life in the home. Their influence was not so much an intrusion, but a means by which domestic life became a means of introspection and self-understanding. Such self-awareness had profound consequences, being itself an aspect of the forces that drove modernization. Understanding themselves differently, given knowledge of their own unique location, house or garden, suburb or nation, Victorians were led to live differently and work differently, play and love differently as well. It becomes obvious when reading many nineteenth-century texts of popular science and home economy that environmental awareness was formed not by one or the other of these elements alone. Rather, it was furthered by attempts to accommodate them all in one habitable space. This was not an easy place to be in. Uncertainty regarding the precise cause or source of debilitating dampness, the exact contribution of foul odours of putrefaction and general filth of smoke and dust to human ill-health was as much a consequence of knowing the environment as it was a cause to banish these annoyances from the home.

EARTH

From the perspective of the farmer or estate manager the earth was, for many centuries, little more than the ground on which one stood or from which plants grew. In the nineteenth century, however, detailed knowledge of the chemical composition of soil replaced the theory of 'humus'. This entailed a longstanding belief in a material existing in the ground that somehow nourished plants and diminished as they fed off it. It was replaced by a more complex understanding of the organic processes forming the 'vegetable mould' of the field or garden and its contribution to plant nutrition and ultimately human sustenance. The application of new chemical methods to agriculture was initiated in 1840 with the publication of Justus Liebig's *Chemistry and its Applications to Agriculture and Physiology* and well-publicized tests on wheatfields at Rothamsted, Hertfordshire. The latter were aimed at increasing crop yields, being a more broadly scaled experiment than earlier investigations in the forcing-house. Such field tests were particularly important for demonstrating the positive effects on plant growth of 'artificial manures' and chemical sprays.[11]

For the home gardener, a more scientific basis for horticultural practices allowed for greater experimentation and the introduction of a broader variety of plant species. Science and technology transformed otherwise barren or impoverished terrain into a palette for reproducing at a smaller scale the abundance and diversity of nature's productions.[12] The city offered unique opportunities for doing so. It was a little-known fact, *Cassell's Household Guide* informed the reader in 1869, that the 'heavily-charged' air of the city was full of ammonia, brought down in large quantities and incorporated into the soil by rains, and was precisely what many vegetables and ornamental plants thrived on. The greater warmth found in metropolitan areas encouraged the growth of certain species, and the best evergreens were to be seen there.[13] Enhanced opportunities required more expert care in manipulating soils and controlling pests and in choosing just the right species for town versus country. In their celebration of the advancements and opportunities wrought by science, such works belied the labour and kinds of discernment that accompanied the use of new discoveries.

Directing attention to the soil in a slightly different way, chemistry recast the earth as the primary resource for animal and human sustenance by providing a basic alphabet for explaining the composition and transformation, consumption and expulsion of matter. After all, as one commentator wrote in 1878, just as 'beef and mutton are only grass and turnips ... in an altered state', it followed that 'the bodies of men are made out of soil'.[14] Soil of a different, equally fundamental kind was valued by the proponents of sewage utilization. Urbanization and the problem of disposing of unprecedented amounts of human waste was turned to advantage by reformers who saw in the immense 'rivers of guano' swirling beneath Britain's cities an opportunity for citizens to repay their 'debt and duty to the soil' by directing their flow to fields and gardens. While chemistry established a common language of fundamental compounds, evident in both the mould of the garden *and* the human waste flowing from the house, the availability of chemical fertilizers justified the growing imperative to dispose of 'night soil' in a less utilitarian, more discreet manner. It was not that soil carried moral value per se, but that the ways in which it was regarded and spoken about and thought about evidenced valuations of different sorts – including moral evaluations.

The enrichment of the soil for gardening purposes and waste disposal were but two of the responses to the many challenges posed by urbanization. As horticulturalists turned their attention to urban landscapes and the needs of the residential garden, soils commonly required extensive treatment. Often they were too 'stiff' or 'clayey' or intermixed with brick and building rubble. Soils were often deemed too 'light' or porous, shallow, too exposed or too shaded by walls and fences.[15] Entailing a related set of concerns, an understanding of the earth's surface strata and formations, its mineral composition and water-carrying capacities, led to a more exact knowledge of building foundations as well. Studies of the load-bearing capacity of various strata and the science of foundations superseded centuries of common-sense observations by builders. Expertise in 'soil statics' was furthered by new large-scale projects such as bridges, tunnels and viaducts and constructions that imposed ever-increasing weights upon the ground of the city.[16] Increased housing densities, the reclamation of wetlands and rubbish yards for building purposes, and the situation of residential areas on difficult terrain required the careful positioning of buildings and gardens. The difficulty of aligning homes relative to the sun's passage was aggravated by needs to maximize drainage, minimize subsidence and prevent the intrusion of 'ground air' from porous strata, coal seams or decomposing organic matter (see Figure 4.2). These were just a few of the 'dangers from beneath', requiring yet further attention to be directed to the earth.[17]

From the vantage point of the home and its surroundings the ground was 'undoubtedly the most unwieldy and ponderous material' given the labour expended by the landscape gardener to level it, form walks on it and keep plants alive in it. From a wider perspective human efforts at manipulating their immediate surroundings appeared 'trifling … when we behold the apparent facility with which nature has arranged it in such a variety of forms'.[18] Knowledge of what lies beneath the soil, including the history of strata and their transformation and the origins of materials that were of use to humankind, could be productive of self-understanding.

Just as urbanization presented the Victorians with particular problems when positioning their homes, such as the importance of aspect and drainage, the popularization of geology gave them reason to consider the broader, moral significance of the earth's habitation (see Figure 4.3). For

T. P. T. Inv.

George Wilkinson geology was the study of the 'wonderful operations of the laws of nature' and of the 'conversion of the rocks of the earth' through human enterprise. In 1845 he attributed a special, moral significance to these actions, as close scrutiny of the earth revealed the progress of culture and civilization. The transformation of elemental nature was the means to transcend the finality of death:

> Without these rocks our habitations and our public work would be mean and perishable; without bridges, canals, viaducts, roads, the result as well as the means of civilisation, we should be unable to stem the torrents of the flood, to exchange the produce of various localities, or to overcome natural obstacles that call forth the energies of man. It is in the mass of marble rocks that the sculptor finds materials to embody the forms of perfectionate nature, and to perpetuate the memory of our illustrious men; and, without the enduring rock, the lofty column and noble temple that adorn our cities, and spread their influence over society, would pass away, transient as the generation that raised them.[19]

In Wilkinson's view the transformation of nature and the durability of stone, marble and granite made human activity appear by contrast all the more fleeting. It became more poignant and worthy of commemoration with those very same materials (see Figure 4.4). Six years after the appearance of *Practical Geology and Ancient Architecture*, John Ruskin published the first volume of his landmark work, *The Stones of Venice* (1851). Both this and Ruskin's earlier book *The Seven Lamps of Architecture* (1849) extolled the use of stone as expressing the 'moral temper' of the people by whom and for whom it was produced.[20] Between efforts to popularize science in the home and the aesthete's turn to geology – the immediacy of the soil in the garden and the more abstract lessons in taste to be prised from the ground – Victorian householders and their builders were cast onto a landscape of vital energies

4.2 'Terrace of the Future on Refuse of the Past', in Thomas Pridgin Teale, *Dangers to Health: a Pictorial Guide to Domestic Sanitary Defects* (1878)

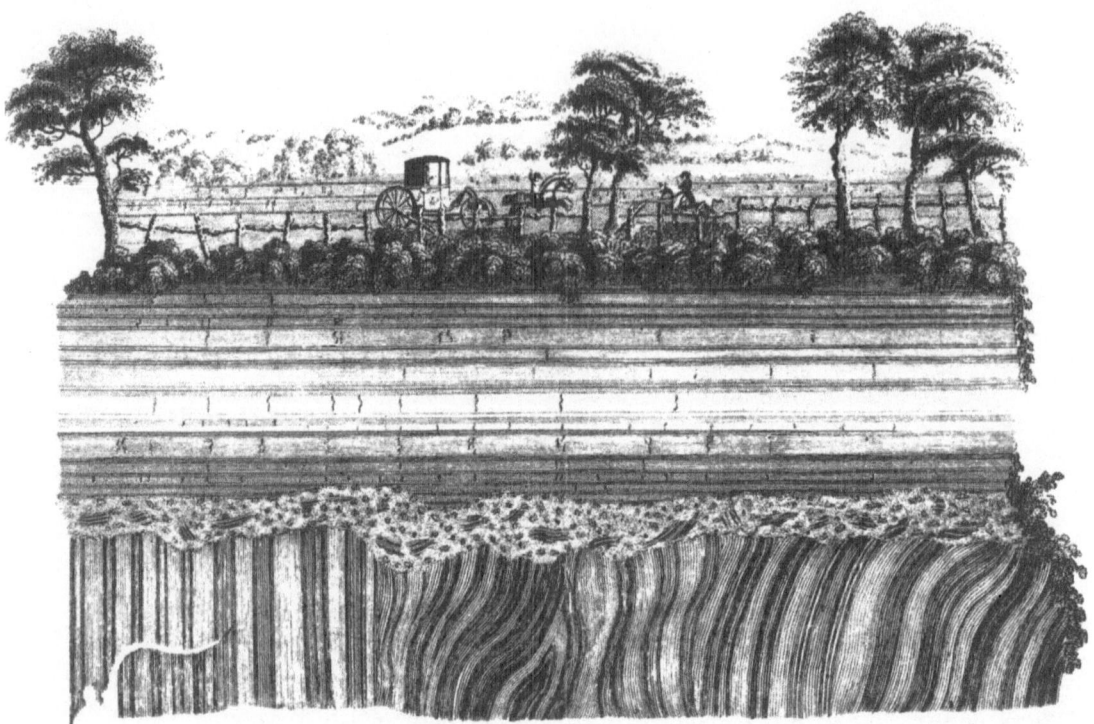

4.3
'Unconformity at Jedburgh, Borders', in James Hutton, *Theory of the Earth* (1795). Hutton was an erudite and wealthy landowner who, like other proponents of earth science in his day, did not allow the practical or economic value of geology to deter him from prising from the earth broader lessons about the place of humankind in the cosmos and its dependence on inanimate nature

to be realized and modes of discernment to be acquired. Their position demanded above all else that the earth be valued and used in certain ways.

AIR

Yet another aspect of nature, the air, moisture-laden, malodorous or dust-filled, was seemingly *everywhere*. As a medium for conducting organic operations, its study furthered a sense of environmental awareness, both close at hand and from a distance. Unlike earth and stone, however, it was perhaps the most indistinguishable of nature's elements. Long before empirical science offered means of assessing the apparent effect of atmospheric temperature, humidity or pressure on living beings, other, older or seemingly outmoded discourses provided metaphors for describing the interdependence of living beings. They invoked the pervasiveness of the air and the contiguity of space itself. The language of machines found its way into the stock of terms the Victorians used to describe their surroundings, and persisted well into the twentieth century. Though not a science of life per se, mechanical metaphors provided a means for thinking about the interconnectedness of rooms and of interior to exterior spaces. In a work such as Robert Kerr's *The Gentleman's House* (1864), the careful arrangement of plans yielded an 'economy' of spaciousness and a compactness in which 'the mind and body, the spirits and even the self-esteem of a man, seem to expand and acquire vigour'.[21] Such a reference is metaphoric in two senses. First, it reflects a desire to reproduce in architectural form the efficiencies of

mechanical devices such as engines or clockworks. Second, it invites consideration that just as air expands to fill a vacuum, so too human physical *and* mental presence is enlarged when entering a room of suitable size. This was a fact of life prefigured by Joseph Wright's crested cockatoo a century earlier, though in a treatise like *The Gentleman's House* it acquired psychological and moral overtones.

More influential than the idea of the 'machine for living', as Le Corbusier would describe the organic character of the home, the persistence of 'miasmatic' theory well into the nineteenth century influenced how air was thought about and encouraged awareness of spaces in the home and the home's surroundings. Initially, the theory invoked the belief that decaying matter released noxious 'exhalations' into the air. It was reconsidered following investigations involving the chemical analysis of decomposition and studies of respiration, though miasmas, humid or rising vapours continued to supply a distinctive set of terms for thinking about the interconnectedness of the organic world. Awareness of miasmas in buildings served to articulate a sense of enclosure by suggesting the sources and paths for the emanation and circulation of gases, the movements of odours and fragrances, breezes and draughts. Outdoors, residents were led to consider the location of their homes relative to the sources of ground exhalations, low or swampy land, places of industry or burial.

With a fuller knowledge of the composition of air – one part oxygen, four parts nitrogen, with a small quantity of carbonic acid and a 'considerable proportion of aqueous vapour', according to one popular account in 1856 – its supposed purity could be conceived of in empirical terms. Similarly, the influence on the residents of spaces in which air was notably *impure* could be

4.4 'The Standing Stones of Stennis', in James Fergusson, *Rude Stone Monuments in all Countries: Their Age and Uses* (1872). Like his broad surveys of architectural history, Fergusson's book reinforced a view that the occupation of the earth by humankind was driven by basic necessities for shelter, against which other cultural values such as the symbolism or purpose of monuments could be assessed

assessed. Manuals on the 'art' of warming and ventilating rooms and buildings were commonly prefaced by the most up-to-date statements of scientific fact concerning interactions between the air and human respiration and well-being:

The action of contaminated confined air has been seen among the most potent and insidious of mortiferous agencies. Any addition to the natural atmosphere that we breathe, must be a deterioration, and absolutely noxious in a greater or less degree; and health … would immediately suffer, did not some vital conservative principle accommodate our functions to circumstance and situation. But this seems to get weaker from exertion. The more we draw on it, the less balance it leaves in our favour.[22]

Mediating between a sense of the body and the rooms it might occupy, between the local and the global, such observations were paralleled by a more complex understanding of 'atmospheric phenomena' as the century advanced. This allowed residents to mediate between local and global concerns. Of increasing interest were the interactions between topography, temperature and air movement, barometric pressure and humidity, and related physical, magnetic and electrical phenomena. These served to identify unique geographic regions according to the climates found there. Within tropical regions, David Ansted related, climate was modified by land and sea breezes which were themselves the result of 'the alternate rarefaction and condensation of air by the heat of the sun in the day, and the cold arising from radiation at night'; but these were 'causes acting only within limited districts'. One had to look for yet other causes, to the 'configuration of the globe, and the manner in which it receives heat'.[23]

Meteorology developed not only through the study of such phenomena, but by questioning relations between broader atmospheric conditions and patterns of organic life. Such relations imparted, among other lessons, a heightened sense of connectedness, for better or worse, between the British Isles and its residents, based on the climate found there, its localities, cities and countryside.[24] Sensitivity to the diversity of these relations, in particular where human beings were concerned, was such that the term 'atmosphere', when applied to a particular room or garden, came to suggest in the nineteenth century the mental, moral or psychological distinctiveness of a place. There too, uncertainty reigned, for the atmosphere of a place, like its 'character' – a topic of considerable interest for nineteenth-century architectural theorists – was largely a matter of subjective value.

WATER

The composition and behaviour of water was a subject of some interest in the seventeenth and eighteenth centuries for reasons both practical and philosophical. Improvements to horticultural practices followed studies of hydraulics, leading to proposals for new means of 'raising' water in wells and for conducting it from afar. Industrial expansion was furthered by new inventions for capturing the force of rivers and driving mills and forges. This involved not only mechanics, being the study of machines, but early forms of fluid dynamics, concerning the action of water upon surfaces or within confined spaces. Of more abstract concern, Joseph Priestley's study of the nature of gas was accompanied by investigations in the second half of the eighteenth century into the composition of water and its apparent conversion into air. Studies of boiling and freezing water were furthered by the use of thermometers, just as these instruments made the heated air in forcing-houses a matter of concern. Advances in molecular theory explained capillary action, the 'seeming spontaneous ascent of water' up small tubes and parallel plates. Chemical analyses eventually accounted for observations of such phenomena as the electrolytic power of water explaining, among other things, the corrosion of copper linings on ships and the mineral content of springs. Various studies informed efforts to extract potable or 'sweet' water from the earth and sea, even from the condensate of steam engines.

Given the persistent need for water for varied purposes, for sustenance and sanitation, fire-fighting or industrial use, its provision was historically the 'first urban problem'.[25] Concerns for water in the home early in the nineteenth century were conditioned by circumstances governing its provision in towns and cities across Britain much earlier. Piped supplies were available to only a small minority of towns, while municipal services were unheard of before the 1830s. Prior to the ensuing period of sanitary reform and battles by government and reformers to wrest control of the utility from private concerns, water was drawn by most from common wells, ponds and reservoirs, streams and rivers and supplies of rainwater.[26] This situation proved increasingly tenuous, due not only to increased demand but to the destruction or pollution of many traditional sources. Both were furthered by the expansion of industrial and building activity, the establishment of abattoirs, tanneries, bleach and dye works along river banks, or the obliteration of springs and ponds. In response, the Thames Improvement Company, established in 1836, was one of several such agencies throughout Britain concerned to prohibit the 'immense mass of animal and vegetable matters' which were incessantly poured into the nation's rivers.[27]

In the home, appreciation of greater or lesser quantities of water in the ground and air made people ever more attentive to the fabric of the built environment, particularly its divisions, junctions and seams. Drainage had long been a key concern for builders. However, the permeability of soil and its capacities to absorb, percolate and potentially purify rainwater, surface and roof runoff were more often than not exceeded in residential districts, which became much more extensive during the Victorian era. There followed the need for storm drains and gutters, devices and traps of all sorts to remove overflow and to limit the back draught of sewer gas. The rusting of iron pipes buried in the ground and architectural fittings in the open air and the corrosive effects of galvanic action between flashings and fittings of dissimilar metals provided a constant source of worry. Moisture posed a threat to the integrity of buildings and their builders as great, though less obvious, as the use of artificial stone. As a greater range of materials and prefabricated building components became common by the middle decades of the nineteenth century, their integration into an overall strategy for water- and damp-proofing the home was a chief concern of the conscientious architect and builder.

Conscious of where water came from and how it flowed and what it might contain, the householder was compelled to relate to the surrounding landscape in various ways. On a broad level, the 'elasticity' of air and its condensation into clouds and rainfall drew attention to the patterns of weather connecting nature's departments. Familiarity with the role that each played in the 'economy of nature' made one ever sensitive to the activities heralded by the arrival of 'rains, perennial springs and streams, the trade winds and monsoons, the regularity of the seasons, spring time and harvest'.[28] The eventual full servicing of houses with piped water made its ready availability a matter of common expectation. Piped water provided a measure of consumption along with health and social progress. Rates for human consumption were known in Britain and America relatively late in the nineteenth century, and they continued to vary and generally to increase.[29]

On a more intimate level, since Anthony van Leewenhoeck's famous observations of *animalcula* or 'living atoms' in rain-, well-, sea- and snow-water in 1676, it was well known that there were other, potentially harmful worlds inside the smallest of raindrops, flower pots and puddles.[30] Along with a menagerie of microscopic creatures waiting to be consumed, the presence of 'putrescent organic impurities', the cause of grave 'derangements of the human system', were particularly frightening. They were generally undetected by the senses of smell and taste, unlike similar impurities in articles of food.[31] As water became seen as the source of a number of diseases and as its consumption rose, the control of water supply in the home proved ever more crucial. After all, as *Cassell's Household Guide* advised: 'Water enters into the composition of all our food, it is the chief ingredient in all our drinks, and it is largely present in the air we breathe. Its absence for a short time only would be followed by the extinction of our very life.'[32] While the idea of 'pure' water may have been ancient, it was given a name by British and French scientists at the end of the eighteenth century, and subsequently became an important product of industry.[33] Following investigations in the 1780s involving the synthesis of water from its constituent gases and then the elaboration of John Dalton's chemical atomic theory two decades later, pure water became simply H_2O. The environmentalist tenor of thoughts on disease and its causes during the Victorian era followed the extension of the site of sanitary practices from dung-heaps and animal sties, vaults and cellars, cesspools, drains and sewers to the water supply systems of entire residential districts and cities. Though works of popular science and home economy informed the homeowner that water was found in practically everything – including a very large percentage of their own bodies, it turned out – this fact did not relieve them of the need for yet further discrimination. Discrimination also involved a need to distinguish between separate kinds of things, whether food, drink or living bodies, which their 'non-water' parts were crucial in constituting.[34] In other words, by imagining water – like air –

in a singular, abstract and pure form, one could conceive of other substances that were less than ideal, impure or potentially threatening in the home.

Dampness, putrefaction and filth

For residents of British cities the 'departments of nature' were hardly distinct, while the effects of earth, air and water on living bodies were notably both positive and negative. Valued for its fertility and stability, on the one hand, the earth could become a precarious, even deadly, terrain, on the other. It was a source of mud and household dust or, worse, a carrier of germs, poisons and other potentially harmful agents. Varied and often contradictory theories prevailing over the course of the nineteenth century cast both soil and water as sources of contagion or infection, while boundaries between these elements were themselves uncertain. The permutation of subsoil by both salutary and noxious forces rendered the air arising from it particularly complex and, significantly, noticeable. Environmental awareness was coupled to new or heightened kinds of sensations, obliging residents to consider their own position relative to animate and inanimate matter, as the one decomposed into or drew sustenance from the other.[35] Odours and other stimuli offered the inhabitant 'some idea of things without', as Locke would describe the world outside the camera obscura, though environmental understanding was not simply a matter of being fixed in place like a lone figure in a darkened room or an insensible plant within a forcing-house. Rather, it required individuals consciously to position themselves within urban and domestic spaces. Often it required them to retreat from the source of unpleasant sensations into rooms or neighbourhood districts purposefully and 'wholly shut' from their influence.

Seeming to possess a power all of its own and to pervade both animate and inanimate matter, dampness left its mark on the novel building forms and building materials of Victorian Britain. Less likely to fire the imagination of visitors to the glasshouse, it was nonetheless a persistent annoyance and could prove more costly in the long run than dead palms or the threat of hail. Dampness was measured by the lines of rust that inched further along the iron pillars of glasshouses and by the streaks of moisture down window glass in all sorts of houses. This was particularly the case during periods of high humidity or winters requiring artificial heat which, respective to each situation, was almost always. The sensation of dampness reinforced awareness of the qualities of enclosure and exposure exhibited by rooms and passageways, even courtyards, lawns and parterres. Lending the authority of science to a popular belief in the harmful effects of humidity within confined spaces, James Cameron, a medical officer for the city of Leeds, wrote in 1892 that:

A damp house has a naturally depressing effect upon the vitality of its inmates. A very damp basement favours the growth of minute parasitic fungi. Malarial diseases, as is well known, are chiefly prevalent in low-lying neighbourhoods, and it has even been suggested that cancer causes more deaths in districts where the subsoil is liable at times to become surcharged with water, than where the site is elevated and dry.[36]

While dampness was made obvious in the home by visible phenomena such as the growth of mould on walls, furnishings and clothing or by condensation on windows, its presence was commonly felt on the skin or, as popular wisdom asserted, 'in one's bones'. A state of mouldiness was cultivated through a heightened sense of smell, emphasizing the important role played by this faculty in forming reactions to the urbanized landscapes of nineteenth-century Europe.[37]

As with dampness, gases associated with putrefaction, whether arising from sewers and other channels of human waste or occurring in the soil, provided all-pervasive and palpable reminders of the threat of disease. They came to be regarded as forming *the* miasma of the late nineteenth century: the source of virtually every communicable illness, with perhaps the exception of smallpox. Traversed by rivers of human waste, the metropolis proved to be the barometer of population growth and weather, as few could afford to escape the city's 'great stink'.[38]

In Esther Copley's *Catechism of Domestic Economy* (1851) the sources of decay and ill-health around the house were numerous, being 'whatever occasions bad smells, or in any way vitiates the air, such as stagnant ponds or ditches, privies, dung-heaps, or the neighbourhood of manufactures or trades in which noxious substances are prepared or employed'.[39] Houses built over common sewers or in low-lying areas were particularly vulnerable in the defensive space of public sanitation and hygiene. Privies and holes in the floor not only allowed sewer gases to enter, but rats – sometimes 'twenty at a time' – forcing residents to maintain a constant vigil.[40] Sewer gases were not only most prevalent in low-lying areas, but often arose near sources of drinking water – increasing fears for public health, as Thomas Teale's 1879 *Guide to Domestic Sanitary Defects* graphically illustrates (see Figure 4.5). Warning his readers that they might inadvertently be drinking their own waste, Teale argued for 'complete disconnection' between the house and the drainage system as the basis of sanitation.[41] His book gave further reason for guarding against ruptures in the fabric of the domestic environment.

Fog and particularly the filth that accompanied it was another intrusion into the living spaces of Britain's cities.[42] Dust, soot and ash not only clung to the surfaces of glasshouses, creating 'messy' kinds of days; they drew the gardener's attention to the cleanliness of the plants in their care, requiring the frequent washing of foliage lest 'impurity' be conveyed through the 'air-vessels' of leaves into the plant itself.[43] In revealing the presence of draughts or by otherwise affecting the lucidity of light indoors, airborne pollutants, often associated with a state of fogginess, engaged yet another form of sensory, primarily visual, acuity. This acuity was nurtured in Britain's new botanical gardens and glasshouses, but also in other 'artificial' environs, such as its exhibition and demonstration halls, libraries and museums, even its railway sheds. Noting the persistence of London's dense and sooty fogs and

T.P.T. Inv.

4.5 'How People Drink Sewage – No. 1. Drain Leaking Into a Well', in Thomas Pridgin Teale, *Dangers to Health: a Pictorial Guide to Domestic Sanitary Defects* (1878)

their infiltration of these spaces, a contributor to *The Builder* in 1865 described the effect of poor atmospheric conditions on life in the metropolis. For the 'bookworms' inhabiting Sidney Smirke's great Reading Room of the British Museum completed eight years earlier, adequate light for reading was as rare as the volumes they sought to peruse:

Many of this industrious and useful community leave the beautiful dome, which in this light has a dim, lurid, and somewhat ghastly appearance, and grope their way homeward; a few, more persevering than others, having with difficulty managed to extract the names and particulars of a few books from the catalogue; and in the hope of the air clearing up, sit with patience, waiting until the painstaking attendants have, in the colossal space of the King's Library, or some of the other mighty lines of book-shelves, by the aid of lanterns, carefully locked and strongly protected with glass or crystal, provided the volumes wanted.[44]

It was no small matter keeping the fog and the dangers it carried out of London's museums, churches, music-halls, theatres and hospitals. Smoke Abatement Societies appeared in many British cities with just that aim. Whether this move was but one response to genuinely deteriorating atmospheric conditions in the second half of the century or was itself provoked by a novel and widespread intolerance for pollution and a heightened desire for air clarity is difficult to say. The concept of pure air, like pure water, was inseparable from the awareness of residents that nature's elements were likely to be contaminated by being present in the city.

Despite the great inconvenience of Britain's weather, the intrusion of fog and filth in both greater and lesser buildings proved instrumental in revealing their permeability to the outdoors. In Westminster Abbey, it was claimed, the fog did not seem to penetrate the interior as rapidly as in St Paul's Cathedral. Its intrusion into the upper recesses of the latter, however, seemed to have had a more injurious effect in hastening the cathedral's decay. Today, a building conservator would be equally concerned about the effects of acid rain. By 1887, the situation had worsened to such a degree at the Houses of Parliament that 'fog-filters' of cotton wool were proposed, so that the Government could continue its business in an appropriately transparent manner, untainted, at least by corruption of an inorganic kind.[45]

In the home the hammer and nail proved as useful as thread and needle, given the homemaker's encouragement to 'fortify' spaces against draughts and the infiltration of dust through the many crevices around ill-fitting windows and doors. This problem was particularly acute because of the slap-dash quality of many residences erected by unscrupulous builders and real-estate speculators. Felt carpet was used for weather strips as well as oil-cloth and twilled bindings, and sheepskin and cloth cuttings formed mats to prevent filth from entering below doors.[46] Filth could make its way into the home through other than atmospheric means. The promise of blissful heat and the potential for housekeeping hell were both carried into the coal cellar, which became an indispensable component of the Victorian house. The cellar had to be located carefully to enable fuel to be conveyed directly from the street, 'without filling the house with dust, begriming everything, and causing endless confusion'.[47] Dust and soot and their effect on human sensibilities were commonly described in such terms. Linking images of the 'sinister caverns' of Britain's coal-mining industry with the abodes of those who were to benefit from its by-products was an 'anti-aesthetic' whereby beauty was countered by chaos and confusion.[48] A heightened sensitivity to the air itself not only gave people yet another reason to hate the English weather or to appreciate the novelty of 'atmosphere control', the forerunner to modern air-conditioning, indoors. It strengthened the link that was gradually formed over the course of the nineteenth century between environmental wholesomeness on the one hand, and physical, mental and social well-being on the other.

Everyday anxieties

Anxieties generated by the unknown composition of air, as well as earth and water, their manner of interaction and responses to sensations of dampness, putrefaction and filth came together in response to the threat of disease. Twenty years before Johnson published *Our House and Garden*, Robert Meikleham surveyed the home not with wonder, but trepidation. Like many writers with a faith in science, he voiced his concerns for the unhealthy prospect of residences with a familiar language for discerning causes and their effects:

It has been remarked that most diseases in all countries have external, and for the greater part *removable* causes; but these being familiar to us are, in the common bustle of life, overlooked or undervalued, and though all may regard them as injurious to their comfort, few will believe they have power to ruin their constitutions.[49]

The threat of cholera was less easily removed from the home and proved particularly ruinous given the great epidemics that swept through European cities during the first half of the century.[50] Searching for a reason for its devastating toll, one writer observed in 1833 that the disease's advance 'resembled the advance of an invading army, and amidst the general consternation which it creates, the cry is "What is it? – is it contagious? – is it an epidemic? or is it malaria? and, what is contagion, epidemic, and malaria? – what are they?"'[51]

In facing the outbreaks of the 1830s, physicians in Britain and America believed that cholera was not a precise entity or active agent added to living systems. Rather it was a mutable and dynamic condition. Its aetiology and symptoms were affected by various climatic and hygienic conditions that could nonetheless be aggravated by the mental and moral state of the victim.[52] As the course of the disease usually involved sudden and widely scattered outbreaks of cases, it was generally believed that the source of cholera was not contagious. Rather its source lay in atmospheric disturbances, attributed to either a change in the normal constituents of the air or the admission of some poisonous substance of terrestrial origin.[53] Commentators warned of the harmful effects of atmospheric electricity, diminished oxygen levels in the air or even the movement of extra-terrestrial bodies such as meteors. In searching for some substance as the root cause of cholera, some even proposed the existence of *animalculae*, or small winged insects invisible to the naked eye.

By the middle of the century, developments in chemistry and biology, clinical medicine and pathology led to new theories that challenged both the atmospheric origins of cholera and the belief in predisposing causes like the inherent immorality of the diseased. These scientific advances directed attention to the possible influence of dirt or decayed matter, lack of ventilation or pure water as contributing factors.[54] Sensitivity to sewer gas and other emanations reinforced the filth theory of disease promulgated in the 1840s and led to the ideas on sanitation of Edwin Chadwick and Southwood Smith.[55] Reasoning that the source of disease could be attributed primarily to environmental factors that arose spontaneously, rather than a contagion or epidemic that was transmitted through human contact or vermin, the English clinician Charles Murchison offered his 'pythogenetic theory' in 1862 that related disease to corrupt organic matter or filth. This was complemented by the German scientist Max von Pettenkofer's studies in the 1850s of environmental pathology and his theory that disease may be caused by poisons developing deep in the ground and their escape into the atmosphere through 'ground exhalations'.

By the 1860s, despite the fact that no chemical or microscopic analysis had as yet identified the precise agent of infection, the very rapid progress of the disease and its extreme toxicity implied that cholera possessed the power of reproduction. It was organic like the beings it attacked, and of a different order to those minerals which, in combination with an acid or oxygen, produce a toxic substance. Cholera, along with smallpox, measles, scarlet fever and whooping cough, joined the list of 'hidden mysteries' lying in wait in the home and garden. Friedrich von Schlegel was to write in his *Philosophy of Life* (1828):

That the air and atmosphere of our globe is in the highest degree full of life, I may, I think, take here for granted, and generally admitted. It is however, of a mixed kind and quality; combining the refreshing breath of spring with the parching simooms [winds] of the desert, and where the healthy odours fluctuate in chaotic struggle with the most deadly vapours. What else in general *is the wide-spread and spreading pestilence*, but a living propagation of foulness, corruption, and death? Are not many poisons, especially animal poisons, in a true sense, living forces?[56]

The study of disease heightened awareness of society along with attentiveness to the conditions in which disease thrived. A kind of social analysis whereby 'moral contagion' – engendered by the particular type of environment in which the poor lived and bred and not simply their own innate constitution per se – was likened to epidemics.[57] Studies of 'mental hygiene' sought to preserve the 'health of the mind against all the incidents and influences calculated to deteriorate its qualities, impair its energies or derange it movements'. Among the physical agents that affected the mind's vigour, none was considered more worthy of attention than the air, according to the American Isaac Ray. Of particular concern was its temperature and movement or other, meteorological disturbances.[58] Sensitivity to the atmosphere of particular localities informed the terms used to describe the unhealthy environs at the centre of Britain's cities. They were frequently derided as 'colonies' or 'rookeries' as though their inhabitants were not fully settled or, worse, sub-human. Concerns were frequently expressed over the state of such districts as enclaves of disease, vice and immorality in the otherwise fluid space of social hygiene. In that many of the more noticeable districts developed around industrial or derelict lands, tanneries, docklands and burial grounds, these attracted particular fears.

Responses to these closed and pestilential places varied in scale, but were remarkably similar in purpose. Public parks provided a context for new customs and forms of social interaction, all the while ameliorating the more obvious, negative effects of industrialization and urbanization. They provided, as it were, means by which society could be imagined as having an obvious or corporeal presence, just as varied references to the soil, atmosphere and water and their improvement were measures of human enterprise and environmental responsibility. Street clearance schemes and model dwellings served techniques of reform. Measures aimed to expose the darker regions of society to observation and regulation were intended to introduce the beneficial effects of civilization along with light and fresh breezes. They were meant to instil an atmosphere of decency and morality as well as health. Equally, the park or botanical garden, the house and garden, the study or parlour were enlarged or improved in light of this heightened environmental sensitivity. There followed greater attentiveness to the boundaries formed between these spaces and nature's basic elements.

The popularization of science in the Victorian home was accompanied by a belief that science was of immense benefit to residents by revealing the origins and laws of their natural condition. To be scientifically minded meant to question how the 'departments of nature' not only administered to each other in a 'marvellous manner' but also how they might provide resources for improving health and wholesomeness of character and thus reinforce the physical and moral temper of the British nation. Self-improvement was implicit in the desire for a more direct communion with nature or, in some instances, its literal 'consumption'. Health reformers and supporters of vegetarianism in Britain and America, notably during the second half of the nineteenth century, renewed a longstanding belief whereby human beings were seen to have degenerated from a state of accord with nature, a state in which a vegetable diet and physical vigour were symptomatic of an ancient or pre-industrial golden age. Disturbed by the seeming artificiality of modern life, they cultivated the vegetable as a part of an idealized natural environment in which 'undefiled instincts directed people's living habits'.[59] The reformer William Alcott argued that a vegetable diet and proper hygiene would bring about not merely a return to nature, but 'a gradual ascent to nature' through a more thorough, scientific understanding of nutrition.[60] In this regard the body was an inextricable part of the broader organic world, governed by the same chemical and physical operations and fired by the same vital forces transforming matter. After all, food, asserts *Cassell's Household Guide*, is fuel:

We require food frequently for just the same reason that a fire requires coals frequently, and a lamp, oil – because we are burning away. Strange as this may appear, it is a most certain fact. The air that we breathe into our lungs contains oxygen, and this oxygen combines with or burns the muscles and other organs of our bodies just as it does the coals, in a fire ... All this heat comes from the slow wasting or burning of the substance of the body, so that it is evident that if we did not make up for this constant loss by eating food, our organs would soon be wasted away and consumed.[61]

Modern views on the subject of diet seem to have changed very little, particularly given the moral defence of vegetarianism and the common emphasis on the caloric value of various foods. Efforts at dietary reform and the regulation of food production during the Victorian era were to influence the spaces and practices of domestic life. The discovery of the benefits of refrigeration in the middle of the century and its gradual introduction in the home was to have a great impact on eating habits.[62] Apart from improvements to the quality and longevity of food, ice-houses gradually disappeared from upper-class houses as supplies were no longer as closely tied to seasonal availability. With the extension of rail and shipping networks and, by the end of the nineteenth century, the introduction of frozen meat from Australia and Canada, dining habits could truly encompass a global scale.

In the garden, concerns for pure or unadulterated food and a focus on the superior nutritional value of vegetables were complemented by aesthetic interests that encouraged the planting of edible foodstuffs. A nostalgia for old-fashioned gardens associated with life in pre-industrial Britain encouraged the mixed planting of edible and decorative species. George Eliot bemoaned the virtual extinction of such traditional gardens, while the narrator in *Scenes of Clerical Life* (1858) described them as mere fragments of childhood memories. There, no border separated the flower from the kitchen garden, 'no monotony of enjoyment of one sense to the exclusion of another; but a charming paradisical mingling of all that was pleasant to the eye and good for food'.[63] In a wonderful scene of organic vitality that promised both visual and edible treats, 'the crimson of a carnation was carried out in the lurking crimson of the neighbouring strawberry-beds: you gathered a moss-rose one moment and a bunch of currants the next; you were in a delicious fluctuation between the scent of jasmine and the juice of gooseberries'.

Where these scents were reproduced in the urbanized landscapes of the private garden, they belied the chemical fertilizers that increasingly were used to produce them. They competed for attention with the stink of compounds used to rid petals and berries of insect pests. Eliot's vision was perhaps coloured by growing awareness of the deplorable state of Britain's food industries. In the first half of the nineteenth century it was a common enough practice to adulterate food, while relatively little thought was given to keeping it fresh.[64] Bread was whitened with alum, flour and milk with plaster of Paris and chalk, while cider and wine were served with quantities of lead. Up until the last decades of the century, efforts to improve milk sanitation remained directed towards restricting the admission of all adulterating materials and the improvement of overall sanitation. Knowledge of milk-borne disease awaited the elaboration of germ theory, one of the most significant advances of nineteenth-century biological science.[65]

Fears of the inadvertent consumption of organic or inorganic impurities were directed not only at sources of food and water, but the walls, floors and furnishings of the home itself. The bright colours, paints and wallpapers made available to the home decorator given the expansion of the chemical industry covered up the potentially toxic environs of lead-oxides and compounds of mercury. Among the many 'valuable discoveries in metallurgical chemistry' brought to the notice of the public and scientific communities was the realization that the lead used in water-pipes and other sanitary appliances delivered numerous 'homeopathic doses of metallic poison'. As a consequence even the 'strongest constitutions are gradually broken down by causes which lie beyond the ken of friends; or even medical men of average attainments'.[66]

The vitality of life

The ancient Greek philosopher Empedocles held that all of creation was composed of four basic substances: earth, air, fire and water. They were unchanging and everlasting. Combined by the force of Love and separated by Strife, they produced the complex material that formed the world. This concept dominated Western thought for many centuries until it was challenged by alchemists and the prospect that nature's elements could be transmuted or transformed.[67] Of the four Empedoclean elements, fire has been left out of the survey of elements in this chapter. Perhaps its combustive and stimulating force proved to be more metaphorical than real. In the texts considered so far it denoted less a precise, causal power like the rays of sunlight that 'cooked' the vegetalist's specimens or the breezes that felt fresh on the skin in *Our House and Garden*, but rather the exchange of energy between living beings and their surroundings. Perhaps its force was the one that turned grass and turnips into beef and mutton and its heat the one that came from the 'wasting or burning' of the body and the food that replenished it. In a manner of speaking, the Greek philosopher's fire became the force of life with the advent of modern science, particularly biology. Accompanied by the physical, chemical and statistical study of earth, air and water in the nineteenth century, this vital force was to be diffused throughout the organic world and to course through the living beings, plants and animals, insensible and sentient creatures, that were to localize and consume it, transform and be transformed by it and ultimately perish by it. This force was of particular concern for members of the human species who sought to use it for their own purposes and to come to grips with the fact that they too were governed by it – somehow.

An important part of everyday experience, attentiveness to earth, air and water in the Victorian house and garden transformed the domestic sphere into a unique habitat by the time both 'habitat' and 'ecology' entered the common stock of terms addressing environmental concerns. This was due, to no small degree, to explorations such as Johnson's of the hidden mysteries of the home. One could argue that books of popular science, domestic architecture, gardening and household economy were each, in effect, an *exposé* on the qualities that the Victorians residing in their houses and gardens *should* have. These books suggested that residents were susceptible to reason, open to curiosity and a sense of purpose in making the home a better place. A common manner of reasoning – of distinguishing causes from effects – pervades many of these works. Acquiring this faculty for drawing meaning from one's surroundings was tantamount to the exercise of foresight, discernment and prudence.

It can be said that of the many forms of advice given to householders during the Victorian era, perhaps none required them to be as environmentally aware – and reasonable – as the guidance offered to prospective home-buyers. House-hunting required the ability to situate oneself within environs, present, past *and* future – to adapt to the potential threat of domestic disaster. Acquiring a home became as symbolically charged, as loaded with overtones of mental and moral prowess, in Victorian Britain as recreational tiger-hunting was in Imperial India. It reflected both a booming market in middle-class homes *and* the value placed on domestic accord. Entering the market for a place to retire from the rigours of life required awareness of the needs and resources that retirement entailed. It required a faculty for distinguishing between one and the other and for foreseeing the consequences of having too many needs and too few resources. Entering the house market was like entering a jungle where the homebuyer encountered Rousseau's noble savage coming the other way, emerging from its 'state of nature' and lack of conscious need. Apart from financial considerations, *Cassell's Household Guide* advised the prospective homebuyer to seek a residence

which does not lie low, or on soil which is at all swampy and ill-drained, and we should try to obtain a house built upon gravel, sand, chalk, or rock. We must also aim at having a good share of sunshine, and light, and air. Even if we choose our home in the suburbs, we shall be wise to look out for an open situation, and neither too closely hemmed in by trees, nor standing upon a bad soil. A very large number of speculative builders will remove from the ground they build every particle of gravel, or other useful subsoil, in the neighbourhood of London, and will have the excavations filled in with all kinds of refuse and rubbish. It is needless to say that one's house might as well stand in a marsh as upon such materials, for unwholesome exhalations will arise, and various forms of disease be induced in consequence. This is not all: the houses erected upon such ground are liable to be damp, and apt to settle down, causing cracks in the walls and partitions, bulging out in some places, and shrinking in others; hence windows and doors get out of order and do not shut and open properly, and expose the inhabitants to draughts.[68]

Science introduced the homebuyer to the dire consequences that awaited the unwary. Libraries and booksellers were well supplied in the nineteenth century with household guides and titles similar to *Our House and Garden*, where appeals to science served moral purposes. By establishing common or everyday phenomena as worthy of study, science transformed the house and garden into places for promoting domesticity as a way of life and provided means for observing the inadequacies of both the home and its inhabitants. Retirement within rooms and passageways and behind garden walls served to extend practices of observation and measurement associated with scientific enterprise and cast the householder and homemaker as the attentive respondents to nature's demand for order and regularity. Traversed by innumerable organic and physical operations, the home was as important an analogue of nature as any glasshouse could ever be. Equally, the home was an important site where concerns for environs of various types and scales were negotiated. The 'sanitarian' Benjamin Richard wrote in 1885:

If in the centres called home the foundations of the science of health are laid, the rest, on a larger scale, will necessarily follow, for the same rule that applies to the accumulation of wealth applies equally to the accumulation of health. 'Take care of the pennies,' says the Financier, 'the pounds will take care of themselves.' 'Take care of the houses,' says the sanitarian; 'the towns will take care of themselves.'[69]

Appeals to science incorporated desires for greater understanding, truth and objectivity. They also begged efforts aimed at discovering one or the other class of organic phenomena or 'abounding marvels' in the domestic sphere, so that these could be turned to human purposes. These efforts required leaps of the imagination to see, in fact, what authors saw when leaving the house and upon entering the garden or vice versa. They promised expertise and a sense of self-satisfaction in controlling the domestic environment, though, with hindsight, they offer some hint of a profound sense of anxiety that went along with the effort.

Patterns on the landscape

Attentiveness to earth, air and water in the home supported the view that the environment comprised numerous organic, chemical and physical phenomena. It was a domain that provoked considerable curiosity – though frequently anxiety – regarding the dependencies arising between living things and inanimate matter involving the causal subordination of one to the other. Well-cultivated 'vegetable mould' caused flowers and shrubs to thrive in the garden, for instance, though humans suffered when mould of a different kind appeared on interior walls and furnishings. Spaciousness imbued mind and body with vigour according to some experts, though only if the atmosphere of a room was free of dust and grime. If not, then even the largest and noblest of rooms amounted to little in terms of health and well-being. While books on domestic architecture and gardening and household economy alerted readers to the reasons for such phenomena, the living conditions they engendered were ultimately caused by human negligence, given the moralizing tone of many manuals. The Victorian home was a means for moral and social improvement as it provoked residents to *think a lot*, particularly about their surroundings and their responsibility for them.

This chapter develops further the idea that environmental awareness was partly an outcome of life in the home, given the expansion of British cities in the nineteenth century. Awareness grew from the popularization of science, but also efforts to design the house and garden to conform to, not simply mimic, the way nature worked. In this, residents allowed for both nature's complexity and its appearance as a unified whole. Their efforts were aimed at providing as suitable a habitat for themselves as the earth's geographic and climatic regions seemed to provide homes for other species and their distinctive ways of life. Allowing for nature's elements, forces and dependencies in the house and garden, residents were compelled to consider how they all came together in a coherent, organized way. Domestic life *was* an economy after all, as many authors and book titles asserted, meaning that the home required balance, vigilance and constant maintenance like a well-oiled machine. Each of nature's departments and every room in the house was related to another, while the whole was greater than the sum of its parts.

This important belief in the organic character of the home and its unique suitability for its occupants was supported by works of erudite theory and nineteenth-century fiction. A sense of the connectedness of interior and exterior spaces was as much a consequence of what residents read as how they planned them. Various literary genres encouraged a particular way of 'reading' domestic environs, of interpreting or making sense of house and garden. Literature encouraged awareness of the environment, promising with scientific certainties and entertaining narratives that there were 'deeply interesting phenomena' about. Genres offered practical advice and proposed unique places to build and live in and ultimately assume responsibility for.

The depiction of place forms a long tradition in English literature. Unique settings become meaningful not simply in relation to other 'real' locations they may represent, but in relation to literary traditions themselves. For instance the country manor or 'big house' provides a key location for many fictional narratives even though actual estates may not themselves be depicted and while the social and political significance attributed to this type of building have varied.[1] There is a shared background of assumed meanings or references connecting fictional and non-fictional works in describing settings where significant or interesting events happen. Nineteenth-century novels and adventure stories, travelogues and journal accounts made these settings 'real', as it were. They provided opportunities to imagine landscapes otherwise experienced partially and close at hand – an isolated view from a window, say, or a house next door – or from afar – surveyed on the pages of atlases or across distant geologic eras. Literature promised new, potentially habitable places with their own distinct geomorphic or organic character, natural forms and living species, paths of approach and ways of passage. Fictional narratives of voyages to distant and exotic lands complemented illustrations of urban jungles closer to home. Gustave Doré's *London: A Pilgrimage* (1872) dramatized the artist's tour of the metropolis with now iconic images of Victorian grandeur and squalor, scenes characterized by the people living and working in them and by particular ambient qualities (see Figure 5.1).[2]

In some literary works the close scrutiny of nature associated with empirical science was itself dramatized. Doré's portrayal of a city struggling with an excess of humanity is mirrored – in reverse – by Richard Jefferies's *After London, or, Wild England* (1885), in which mysteriously de-populated towns and countryside are found languishing, suffocating under the weight of nature's fecundity and uncontrolled floral growth. Works such as *After London, or, Wild England, The Island of Doctor Moreau* by H. G. Wells (1896), or in the twentieth century *The Lost World* by Sir Arthur Conan Doyle (1912), exploited an ensemble of concepts, concerns and anxieties arising from a general openness to issues of ecology and evolutionary science. They offered a view of the past as having been formed by incessant change. It was a view more expertly addressed in the landmark work of 1864, *Man and Nature*, by the American George Perkins Marsh. Heralded by the philosopher, historian and architectural critic Lewis Mumford (1895–1990) as the 'fountainhead of the conservation movement', Marsh was perhaps the first to point out, in a

5.1 Gustave Doré, illustration of Houndsditch at night, *London: A Pilgrimage* (1872)

methodical and comprehensive way, 'the dangers of imprudence and the necessity of caution in all operations which, on a large scale, interfere with the spontaneous arrangements of the organic or the inorganic world'.[3] In works of both theory and fiction curiosity about nature was never far from fears of its neglect.

Literature is not only a medium for representing ideas or describing characters, places and actions. It can also be a means for shaping the moral character of the reader and of society as a whole. The popularity of the preceding fictional works followed calls for the study of English literature in Britain's new public schools during the middle decades of the nineteenth century. This was part of a broader pattern of change in educational thought in several European countries, where national languages gradually came to replace instruction in esoteric Greek or Latin. The idea was revolutionary, one with consequences for individual identity and society comparable to the growth of environmental awareness at the same time. In Britain, by providing innumerable stories drawn from life and presumably close to the experience of even the poor and illiterate, the English language was purposefully promoted by educationalists and reformers as a means of self-discovery and personal development.[4]

Likewise, during the Victorian era reading about unique places was akin to creating them in the home. Home improvement, gardening and interior decoration were important modes of imaginative play, self-expression and discovery, whereby a belief in the spontaneous order of the natural world informed practices aimed at transforming residents into 'better' – that is more responsible, more productive – citizens through the care of their domestic environs. Through its care, the Victorian home became a unique place like the public classroom, though one for studying the environment, being a domain where nature's order was appreciated and put to work. The home was a place for expressing oneself creatively as well as for exercising restraint. It was a place that, when properly designed, should possess gardens that appeared natural and interiors with appropriate ornamentation. To design the home well was to form a dwelling place that other living creatures, were they sensible of such things, would find as comfortable as their own.

Appreciation of life's mysteries in the house and garden furthered a belief in the purposeful order of nature, while efforts to bring nature indoors or recreate it behind garden walls were sought through a number of formal means. Frequently the forms of organic nature were studied and imitated as a way of achieving a closer bond with the natural world, a sense of fitness or adjustment to its principles of order and purpose. This particular view of 'organic' design, whereby the principles of generation and growth governing, say, the configuration of a leaf or flower or seashell were applied to objects of human design and manufacture, is evident in nineteenth-century visual culture. Contemporary appraisals of the influence of biology on nineteenth-century design cite many instances of the organic in architecture. These range widely and frequently include Semper's theory of natural or formal types and Owen Jones's monumental *Grammar of Ornament* (1856) that surveys, amongst other sources of decorative forms, the use of organic motifs for aesthetic purposes. Reference is commonly made to the works of Arts and Crafts practitioners such as William Morris and, at the end of the nineteenth century, the sinuous forms of Art Nouveau. Such episodes have been well rehearsed.

The concept of the 'organic' entails not only historical efforts aimed at imitating natural forms. It also incorporates the broader search by householders and the designers in their employ for formal coherence and, hence, a sense of the 'suitability' of their surroundings. Supervening on such coherence and suitability was a feeling of mastery over them – the creation of a unified spatial experience in the house and garden.

Literary landmarks

In many fictional works of the Victorian era, nature was an essential and integral part of their overall imaginative structure.[5] For those whom the famous geographer Alexander von Humboldt (1769–1859) acknowledged to be 'endowed with susceptibility' for natural beauty, read fiction alongside studies like natural theology, biology (with its attendant disciplines of botany and zoology) and, eventually, evolutionary science. In general terms, these studies described the natural world as a system of dependencies formed between species and between organisms and their surroundings. Thoughts about 'ecosystems', as they came to be known in the twentieth century, and the 'wild places' where they were most clearly observed, were seldom far from thoughts about the dependence of humankind on other living beings and inorganic matter. Consider the comments of Robert Hunt in 1844 and his *Researches on Light* where one finds portrayed the interconnectedness of all living things and a peculiar fascination with colour as a sign of life's vitality and diversity, measurable in degrees. In the following passage there is a sense that the entire globe formed an immense conservatory compared to which iron-and-glass ones were but pale imitations:

Where the influence which accompanies each ray of Light can penetrate, there we find organisation and life. In those abysses to which it cannot reach, is an eternal blankness. Even in the pellucid ocean, we find, at no very considerable depth, where the faint gleam of Light is dying into darkness, a few animals and these few of the lowest order of organisation, and colourless. Below this region, neither vegetable nor animal is found … Even on the surface of the globe the influence of the sun is shown in the most marked manner. The animals and the plants of the tropical climes glow with the richness of colour; those of the arctic regions we find them nearly colourless. The races of men are characteristic of the clime in which they are found, not in the colour merely, but in physical power, in animal passions, and in mental energy.[6]

The mention of races here shows the way this concept of the unity of organic existence was easily exported from the natural sciences to the social and historical sciences as well as to fiction, where it was taken up in issues like character, disposition or temperament. It was but a short step from Hunt's observations to claims that the superiority of the British race rested in part on its culture and institutions, but ultimately grew from history and its unique position on the earth.

John Hutton Balfour (1808–84), the renowned professor and key proponent of 'Taxological' or systemic botany, shared with figures like Humboldt and Hunt an appreciation of the impact of geography, topography and climate on forms of life. He was largely responsible for promoting the distinctiveness of regions on the basis of the distribution and character of living beings found there. Regarding the geographical distribution of plant species he wrote in his seminal text, *Outlines of Botany* (1854):

The nature of vegetation covering the earth varies according to climate and locality. Plants are fitted for different kinds of soil, as well as for different amounts of temperature, light and moisture. From the Poles to the Equator there is a constant variation in the nature of the Flora ... The same thing is observed in the vegetation of lofty mountains at the Equator, in descending from their summit to their base ... Each zone, however, has its own peculiar features.[7]

Balfour challenged natural historians to enlarge their understanding of the organic world by observing it directly, just as Henry Home had appealed to the gentleman farmer decades earlier to breathe the air outside their musty libraries and explore nature first hand. Balfour thought that:

Excursions may be truly said to be the *life* of the botanist. They enable him to study the science practically, by the examination of plants in their living state, and in their native localities; they impress upon his mind the structural and physiological lessons he has received; they exhibit to him the geographical range of species, both as regards latitude and altitude: and with the pursuit of scientific knowledge, they combine that healthful and spirit-stirring recreation which tends materially to aid mental efforts.[8]

As an aid to 'mental efforts' visits to wild places contributed to understanding their significance as habitats for communities of plants and animals. This practice was not only advocated for professional scientists, but was adopted by scores of amateur naturalists who found wilderness at home equally informative (see Figure 5.2). For both professional and amateur, plants remained particularly interesting for they, unlike animals, did not generally move about and could be observed easily and without fear. Study in the field was encouraged by Balfour and British horticulturalists, keen to promote botany as a means of educating the general population into the situation of humanity in the organic world and, hence, its moral condition.

Seeking to inspire respect for the order of creation, Charles Smith argued (1852) that the botanical collection in the newly completed Palm House at Kew should be rearranged according to systemic, 'territorial' or 'divisional' principles:

for, being skilfully executed, they would posses a high degree of interest ... At least three of the quarters of the globe – Europe, Asia, and America – could be represented; and subordinate sections, if necessary, might be made to include the plants of the more important countries or ranges of continent embraced in the main divisions. For example, we might have a British Flora, a French and German Flora, a Mediterranean Flora, a Russian and Siberian, a North American Flora, and various others.[9]

Geography was a means of fixing species, no less so human beings, within distinct regions or nations. In Smith's view, without explicit lessons in geography to inform the spectator of just what they were admiring while amusing themselves at Kew Gardens, its glasshouses were but mirrors for self-indulgent spectacle. They were also, potentially, nothing but vehicles for a kind of botanical eclecticism for, as Smith continued:

Undoubtedly, one of the most curious things in our flower-gardens is to see natives of the Alps and the Himalayas, of Oregon and of the Cape of Good Hope, all growing peacefully and lovingly together; but in consequence of this promiscuous planting, the facts of botanical geography are at once lost sight of, and very often completely forgotten.

5.2 'Group of Cockatoos', in John Wood, *The Illustrated Natural History* (1863)

The prospect of 'promiscuous planting' is a reminder that artifice was not very far from the territories recreated in the public glasshouse and the private garden. Of implicit concern to Smith, but also to curators of zoological collections at the time, was the possibility that exhibits would fail to enlighten the viewing public, becoming rather circuses of exotic forms serving no greater purpose than novelty itself.[10] On the contrary, where nature, plant and animal species were arranged according to geography or systemic principles, opportunities existed to reveal something of humankind's dependence on the environment following an appreciation of the distinctive habitats and ways of life of other species.[11] Equally, exhibits could offer tacit lessons in the usefulness of nature, its availability to human exploitation and control. Continuing his appeal for the organization of plant species in glasshouses according to the regions in which they grew, as opposed to an arrangement governed simply by 'scientific' – meaning here, taxonomic – principles, which were of interest mainly to the professional botanist, Smith argued:

From the divisional garden we are now recommending, persons about to travel might receive at once a general idea of the vegetation they are likely to meet with in foreign countries; and all might learn with little trouble how much we are indebted to particular regions for the trees, shrubs, and plants that tenant our gardens. Certainly the unbotanical public would find more attraction and instruction in these than in strictly scientific arrangements.[12]

By 'little trouble' Smith meant for visitors not gardeners. There were difficulties, he acknowledged, in carrying out 'territorial' principles, particularly 'in regard to those exotics which require protection and artificial heat'. When the second Director of Kew Gardens, the younger Hooker, Joseph, ordered the 'rearrangement' of the plants in the Palm House in 1855 according to the regions they came from, it was with the view to control overgrowth and render it 'more agreeable to the eye and more instructive'. His curator complained of the enormous amount of work involved and bemoaned the loss of specimens to Hooker's enforced pruning.[13]

Both kinds of 'spirit-stirring recreation' in the glasshouse, the visitor's thoughtful perambulations and the gardener's labours, provided a support, not a substitute, for travel abroad. Botanical collections were useful not only in providing information about exotic species, but in accustoming the British to their distant territorial possessions. The organic link established between species and territories that underscored the concept of racial and national identity was to legitimize the Briton's sense of intellectual, cultural and moral superiority in having the ability not only to transform their own land, but to subjugate the flora, fauna and human inhabitants of other places as well. These beings were obviously more distant from home and, therefore, seemingly more distinct from themselves.[14]

In *Phyto-Theology*, Balfour wrote of the fundamental distinction in the world between living beings and inanimate nature, a distinction which is most apparent when comparing their modes of existence:

The ceaseless tendency to *change* manifested in the life of the former, stands in obvious contrast with the unaltering stability of the latter. The snow-capped mountain rears its summit to the clouds comparatively unaffected by the lapse of ages which have rolled by since its first elevation. But what, compared with *its* permanence, is the duration of any structure subject to the conditions of vitality?[15]

From the depths of the 'pellucid' ocean to the summit of snow-capped mountains, from the poles to the equator, from tropical climes to arctic regions, the natural theologian, botanist and evolutionist alike saw the world as composed of species suited to an endless variety of habitats and ways of life. They saw one species, in particular, destined to exercise dominion over all others.

TOPOGRAPHICAL FEATURES

This view of the ordered arrangement of the organic world was reinforced and given narrative form by British writers of the Victorian period and early twentieth century who sustained the imaginative exploration of nature with numerous fictional settings. A few are worth mentioning for they indicate not only what the Victorians were reading, but also the wild places, landscapes and topographical features, islands, rivers and mountainous regions that were commonly represented. The interpretation of images and patterns of natural order found in literature were paralleled by techniques for introducing elements of novelty in the house and garden in a controlled manner.

In 1858 Robert Ballantyne's *Coral Island* appeared, hilly and 'covered almost everywhere with the most beautiful and richly coloured trees, bushes, and shrubs'. Three decades later and many leagues distant, Robert Louis Stevenson's *Treasure Island* (1883) was situated off the coast of Mexico. Though mostly covered in grey woods, punctuated by occasional clumps of pine trees, it possessed foliage which, in places, was 'poisonously' bright. The castaway narrator of H. G. Wells's *Island of Doctor Moreau* (1896) presumed his position to be in the region of latitude 5° 3' south and longitude 101° west, being a small, low island of volcanic nature. Covered with thick vegetation – chiefly a kind of palm, the reader is told – the island's fauna proved much more noteworthy. Edith Nesbit's *Island of the Nine Whirlpools* (1899) was fertile, though remote, being one thousand miles from anywhere. Regardless of where these places were meant to appear, it is worth noting the similarity in their authors' attention to detail and the close description of island forms and environs in Charles Darwin's account of his voyages on the *Beagle* in the years 1832 to 1836. In both kinds of works, one learns of locations, geomorphologies and climates, these islands' indigenous species and human inhabitants (see Figure 5.3).

Rivers and their valleys proved equally serviceable as models of a bountiful and contiguous nature, introducing explorers to new, potentially valuable lands and readers to sites of adventure. Regular reports of the exploration of the Nile or Amazon could prove as thrilling as William Morris's fictional tale of travel on the mighty river in *The Sundering Flood* (1897). Frequently, both real-life explorer and literary protagonist traversed the same topography. To find Henry Rider Haggard's *People of the Mist* (1894) a reader began at the mouth of the Zambezi River in Mozambique, perhaps accompanied by one of many atlases of Africa published at the time. The trial of such journeys is clear in *The Lost World* of Sir Arthur Conan Doyle (1912). Set along the Amazon River, the plateau of the *Lost World* is reached after many trials including skirmishes with mosquitoes and the Jaracaca snake, purportedly the most venomous and aggressive reptile in South America. Having survived such zoological encounters and upon discovering the path through the basaltic or 'plutonic' cliffs surrounding the plateau, characters confront a world of thought-to-be-extinct iguanodons, allosaurs, megalosaurs, sabre-toothed tigers and pterodactyls.

5.3 'Polynesian Island' views, in Charles Darwin, *Journal of Reseaches into the Natural History and Geology of the Countries Visited During the Voyage of H.M.S. 'Beagle' Round the World* (1891)

POLYNESIAN ISLAND.—LAGOON SHAPE, COMPOSED OF CORAL.

POLYNESIAN ISLAND.—CRYSTALLINE STRUCTURE.

POLYNESIAN ISLAND.—VOLCANIC.

Mountains or alpine regions served to identify remoteness and inaccessibility, with the Himalayas, Scandinavia and – not least – the Scottish Highlands serving as popular destinations. Much of Walter Scott's fiction was set in Scotland, reinforcing a tradition where the country served to support various, frequently opposing associations. The Highlands were the scene of rugged wilderness and isolated civilization, a region bearing the weight of history and impact of industrial and agricultural revolution. In 1887, William Henry Hudson drew on the same descriptive and thematic possibilities as Scott by locating Coradine, a fictional land with notable architecture, somewhere in the Highlands.[16] All the buildings there were ancient and inhabited by a people whose origins, like the hills themselves, were 'lost in the mists of time', an allusion perhaps to a land which, historically, had been largely cleared of its residents through the enclosures acts of the mid to late eighteenth century. In fiction, however, the Highlands were seemingly teeming with life.

By and large, the preceding texts used topographical features to structure fictional narratives, subordinating the actions of characters to an imagined experience of novel places. They guided the reader to form their own reactions to these settings by offering an idea of what they looked like, the species found there and the ways characters interacted with their environment.[17] Books on domestic architecture, gardening and home economy invoked narratives of a similar sort. They used familiar features like garden plant beds or furniture arrangements, paths and corridors in a 'topographical' way to articulate an ordered, though varied, spatial experience. Commonly written, as one particular work asserts, for those of 'enquiring mind and eye of taste', such works often advised readers to pursue variety and the 'most interesting things'. This was so, for it seemed many believed at the time, as today, that the imagination as well as the eye 'becomes tired of repetition and sameness'.[18] Along with variety, the desirability that arrangements of species and physical objects, particularly in the garden, should present a 'natural effect' is equally espoused.

ISLAND EDENS

Mention should be made here of the historic background behind the Victorians' appreciation for novel places both at home and abroad. The imaginative exploration of the wild was largely preceded by voyages to real places, some evocative of Eden, others simply strange or exotic. Historically, the desire to locate paradise on earth was at first inspired by scripture and reinforced by Enlightenment philosophers. It was given form by Western responses to early colonial outposts like the Canary Islands, Madeira and Mauritius and, subsequently, European settlements at the Cape of Good Hope and further east.[19] Among various topographical features, islands in particular have long held a prominent, though ambiguous, position in Western thought and literature. As places set apart, removed from the rules and manners that bound civilized society, islands have been seen as somehow more natural than other places. Island narratives are often 'reflections on origins, the site of that contemplation being the uninhabited territory upon which the conditions for a rebirth or a genesis' of European society are made possible.[20]

Discovered over the long period of Britain's colonial expansion, real islands and their fictional counterparts provided opportunities for thinking about the organization of living species within a given, bounded domain. They provided natural historians or biologists with laboratories, as it were, for studying close at hand the mode of living of plants and animals found there. While providing authors with opportunities for imagining humankind in an 'untouched' state of nature, the experience of islands also provided evidence of humankind's ability to contaminate, degrade or destroy the environment and, by extension, the abundance that nature promised. The idea of an island could also be expressed as a morally desolate landscape – where people had to invent morality from scratch – or else used to illustrate a society lawless and without morals. The island was not only a means for recognizing the distinctiveness of a given place; it was also a means for assessing the 'common ground of British imperialism' and, by extension, British science as well. It was a site for both the exercise of knowledge and power and for assessments of individual character and the limits of authority.[21] In this regard the island could represent a microcosm of society.

By the Victorian era, a widespread sensitivity to geography and the systemic or ordered distribution of plant and animal species made one receptive to the characteristics of settings and of the impact these had on the experiences of their occupants. For travellers abroad visits to island Edens were complemented by perambulations through botanical gardens established at colonial outposts. Both were particularly evocative for, as Richard Grove has observed:

The garden organised the unfamiliar in terms of species. The tropical island allowed the experiencing of unfamiliar processes in a heightened sense, both because of the symbolic role which the island was expected to perform and because of the fast rate of geomorphic changes in the tropics.[22]

Enthusiasm for evocative settings was stimulated by visiting or reading about places like these. Reading encouraged ways of interpreting them and modes of self-expression whereby they became inspiring places. The newness of tropical islands and gardens was reinforced by language for describing forms of organic life, habits and ways of life. One could relate to these places given various literary techniques, though perhaps they were not immediately familiar. Techniques included the objectification of scenes into settings for forms of interaction between human characters and lesser species. They encouraged forms of reader identification whereby the portrayal of organic nature, flora and fauna rendered the reactions and sentimental responses of characters understandable. It was a kind of reading that allowed for kinds of *topoi*, spatial motifs or landscape features, to serve as models for distinctive experiences. Islands, along with rivers, mountains and other features, incorporated vantage points and relations of interior and exterior spaces into logical patterns of movement, anticipation and surprise.

LOST WORLDS AND MOLE-HILLS

In visiting botanical gardens at home, the British found novel settings as bounded and packaged for the viewer as those in the accounts of Victorian explorers to distant islands and dark continents. In the botanical garden and public park the vagaries of travel were subsumed by leisurely wanderings along prescribed paths and none-too-subtle acts of self-discovery, given guide books describing what to see. The devout naturalist and glasshouse enthusiast, Philip Gosse, found a powerful tonic for the imagination and an antidote for wanderlust among the 'tribe of palms' occupying Kew and its Palm House:

Reviewing as a whole what we have been looking at in detail, we cannot help being struck with the enlargement of ideas which such a house as this is calculated to give. We can form some approximation to an idea of tropical scenery, far more correct at least than any number of volumes could convey; and of it may, in a subordinate degree, learn what is true of a personal acquaintance with equatorial regions.[23]

Gosse found the 'ample and unexplored field' of discovery that spread before Thomas Knight fifty years earlier. In the glasshouse the environment was offered as an evocative, though ordered and managed, terrain. It was one that inspired the theories of Alexander von Humboldt, who also encouraged artists to visit as well:

He who is endowed with susceptibility for the natural beauties of mountains, streams, and forest scenery, who has wandered through the countries of the torrid zone, and has seen the luxuriant vegetation, not only upon the cultivated shores, but in the vicinity of the snow-capped Andes, the Himalaya mountains, and the Neilgherry hills of the Mysore, or in the wide-spread forests between the Orinoco and the Amazon – that man can alone understand what an immeasurable field for landscape-painting is open between the tropics of both continents, or in the islands of Sumatra, Borneo, and the Philippines, and how the most splendid and spirited works which man's genius has hitherto accomplished cannot be compared with the vastness of the treasures of nature, of which art may, at a future time, avail itself.[24]

This same 'immeasurable field' was an inspiration to home gardeners as well as to scientists and artists. In the bounded world of the suburban garden, however, limits were placed on creative ambitions. In a manual or pattern book of gardening by Joshua Major (1861), intended for 'ladies and their gardeners', for instance, readers were encouraged to introduce elements of horticultural novelty and botanical beauty.[25] They were given permission to combine formal or geometrical flower beds with 'English' ones despite the fashion of earlier times. This was only allowed, however, if a property was of sufficient size to allow for a generous swath of lawn or plantation between them. These were to be provided with a walk 'winding gracefully' through the garden to separate the two bedding systems and preserve their distinctiveness and integrity – their readability, as it were. By this means, one might 'pleasantly pass from one scene to another, enjoying constant variety' without the spectator becoming aware of the full extent of the grounds, its walls and adjoining properties. These, the 'careful Landscape-Gardener is always anxious to conceal'. Likewise, walks were to wind near to flower beds to allow for their close observation, particularly in winter. If they did not, then the lawn would be easily trodden upon and muddy footprints prove as distracting as neighbouring houses. Raised garden beds, the reader was advised, should appear only on undulating lawns to maintain an allusion to hillocks rather than 'mole-casts' rising abruptly upon level ground.

The otherworldliness of distant or exotic places in literature served to heighten a reader's sense of removal from everyday experience and more familiar landscapes, flora and fauna. The depiction of tropical environs conveyed through fictional settings of islands or remote equatorial regions was complemented by territory revealed slowly after long journeys across seemingly endless deserts, or even, as Mary Shelley's *Frankenstein* (1818) illustrates, fields of ice and snow. In their use of narrative to represent a spatial experience genre as diverse as novels as well as reports of scientific expeditions, popular journals and travelogues were equally inspiring as the sermons of any natural theologian or devout botanist. Similarly, Darwin's discoveries were made all the more provocative by the isolation of places like the Galapagos or the islands of the South Seas where he made them. Those of his purported rival Alfred Wallace were made all the more exciting by accounts of his journey along the Amazon River. Numerous readers followed the movements of David Livingstone, who preceded Haggard's fictional exploits on the Zambezi by forty years (see Plate V).

Making apparent aspects of the environment that might otherwise go unrecognized, fiction and non-fictional narratives rendered specific types of places amenable to imitation at home. Literary landmarks furthered associations between places and particular kinds of experiences. In modern parlance, these places became available to psychological projection. As narratives, fiction and non-fictional works made this possible by their very structure, the use of settings and sequences of events, characterization and reader identification. Being obviously rhetorical or intended to persuade, stories led the reader to expect an ordered world where nature was composed for effect. This entailed a landscape of sensibility where forms and boundaries remained somewhat imprecise and uncertain. In literature, this imprecision heightened the symbolic effect of islands, rivers and mountains. Knowing exactly where Doctor Moreau's island or Scott's Coradine was did not really matter as long as they were far away or distant in time. At home, boundaries of particular sorts were required between garden beds, winding paths and undulating lawns to preserve the illusion of limitlessness or the fantasy that one was alone.[26]

When it came to representing the complexity and wholeness of the environment in literature, or when reproducing it at home, the boundary between fact and fiction was equally uncertain. One remained alert to signs of artifice or contrivance as somehow 'unnatural', as evidence of human involvement – some would claim interference – in the spontaneous arrangements of nature. In reading the island adventures of Ballantyne or Stevenson one came to expect a congruence of geographical and botanical detail, 'richly coloured trees, bushes and shrubs' or a tropical verdancy 'poisonously bright'. Conversely, the otherworldliness of *The Lost World* is underscored not only by the remoteness of its setting, but by the certainty that iguanodons no longer exist in the real world except, perhaps, in facsimile (see Figure 5.4). The home gardener might succeed in making a hillock out of a lawn, but never a mountain out of a mole-hill.

Designing by nature

Home ownership in Victorian Britain provided numerous opportunities to arrange and appreciate plant and animal species, in a heightened, symbolic manner. Homeowners were encouraged to observe and sustain organic processes within a circumscribed space and led to position themselves in relation to their animate and inanimate surroundings in a variety of ways. These opportunities demanded powers of discernment and the judicious management of possessions. In contrast to the colonial's experience of Edenic islands, the tightly controlled environment of the home made novelty more obviously an outcome of design. Design brought together the imaginative ordering of space and a thorough knowledge of horticultural science and domestic economy. There followed an emphasis on the overall appearance of landscape features, the arrangement of interiors and the coordination of the one to enhance the other. Generally, attention was paid to subdividing space. In the garden attention was given to the condition of boundaries, borders and garden walls and to efforts limiting or enhancing the impact of seasonal changes along with controlling views. In the house, particular attention was paid to the extent, arrangement and character of rooms and passageways, doors and windows.[27]

The manipulation of landscape features and interior décor in a coordinated and thought-provoking manner was not new to the Victorian period. Nor do the measures mentioned so far – the design of walls and the disguise of the garden's limits, the fashion for bedding systems, thoughts on walkways or furniture arrangements – exhaust the range of skills acquired by Victorian gardeners and homemakers.[28] In previous times, similar practices taxed the imaginative powers of gardeners, architects and literati alike, though generally on aristocratic grounds, palaces, country houses and pleasure parks. Attempts to create an apparent extent of ground on estates, along with accentuating views from rooms within, for instance, were not uncommon.

What *was* new by the second half of the nineteenth century was that the growth of gardening and domestic economy coincided with thoughts by social reformers of the benefits to be had by enhancing the imaginative powers of the British people. Consequently, creativity was encouraged in the home – as in Britain's new public schools – as a means of self-improvement.

The Secondary Island.

1. Mosasurus.	6. The Hylæosaurus.	10 The Ichthyosaurus Communis.	13. The Labyrinthodon Salamandroides.
2 & 3. Pterodactyles.	7. The Megalosaurus.	11. The Ichthyosaurus Platyodon.	14 & 15. Dycynodons.
4 & 5. Iguanodons.	8 & 9. The Teleosauri.	12. Plesiosaurus Macrocephalus.	16 Labyrinthodon Pachygnathus.

5.4 'The Secondary Island', display of pre-historic creatures, including pterodactyls and iguanodons arranged on an island setting in the Crystal Palace, 1851, in Samuel Phillips, *Guide to the Crystal Palace and its Park and Gardens* (1858)

Training in gardening and domestic economy of a considerable portion of Britain's population was legitimated by appeals to fortify the moral, psychological and physical character of the nation's inhabitants by encouraging the exploration of other worlds of human experience. Evolutionary science was increasingly called upon to argue the progressive development of individual sensibilities and social instincts through literature and the arts, independent study and home diversions.

Advocating the humanizing effects of flower gardening, for instance, David Thomson cited in 1868 recent scientific discoveries that revealed the earliest periods of the earth to be verdantly, though monotonously, covered by giant Club-mosses and Ferns.[29] Fragrant, flowering plants appeared only with the 'era of humanity', forming part of humankind's 'preparation in that Eden home' where 'a delicately sensitive human organism and an emotional mind were to vibrate like a well-strung harp of a thousand strings to every influence from without. Likewise, the author of the popular *Book of Home Pets* (1861) advised those who had 'no garden ground to call their own' that numerous plant and animal species were ready servants to satisfy a human need to be close to nature and to learn of its order. Finding 'organisation and life' even beyond the limits and influence of daylight, it was claimed that:

even the dark and dull back court can be made to teem with vegetation of graceful form and pleasing colour, and to prove that a Home Pet of the vegetable kingdom can be established within the realm over which they exercise lordship and mastery – even though it be bounded and encompassed by four walls, the floor, and the ceiling

– which will flourish with all the freshness and brilliancy of colour of its dew-spangled relatives that wave in the lands and hedgerows of the breezy country.[30]

Such claims echoed within the confined spaces and between the walls, floors and ceilings forming the fabric of Britain's expanding domestic environs. Increased material wealth and leisure made gardening and home improvement means for reconciling residents to the prospect of ever more exactingly prescribed living spaces in which creativity and variety were called for. These provided a background of 'inanimate objects', to invoke Balfour's observations of broader creation, against which cosmopolitan life appeared as a series of unfolding events.

CULTIVATING COLOUR

The subdivision of small garden plots, the desire to make them a source of visual interest and the need to care for individual plants species – providing easy access for their trimming, watering, even dusting to remove the 'pollutions which choke their pores' – provoked endless commentary on the extent and arrangement of planting beds. In many yards there appeared archipelagos of oblong, oval, star and half-moon shaped beds set amid grassy lagoons fringed by box-trees and evergreen shrubs, grass or gravelled shores (see Figure 5.5).[31] As John Robson observed in 1860, the widespread adoption of the bedding system of gardening whereby plant species were arranged in groups and patterns changed the nature of gardening entirely.[32] It offered a technique through which design became a means, not merely of ornamental arrangement, but of achieving a purposeful visual and emotive effect in the domestic landscape. Formal, compositional qualities were addressed in light of a more scientific understanding of perception and how human understanding worked to discern differences between organic forms. Perception was guided by a presumably universal desire to impose order on the otherwise chaotic character of unchecked natural growth.

In gardens, attention was given to the size and texture of foliage and how, by varying or contrasting these characteristics, one might further the appearance of order. Colour was a chief means of achieving the overall organization of species and varieties in beds. As the activities of the Maidstone Gardeners' Mutual Improvement Society in 1864 attest, by the middle decades of the century, the interest in colour became as great among horticulturalists as it was among scientists. Such was the case that gardeners kept themselves informed of recent theoretical developments by inviting professional artists to advise them in 1864. Society members recorded the results of each year's colour experiments, interpreting scientific findings and theory in light of shared experience of combinations which, though not entirely conforming to experimental results, were nonetheless 'chaste and pleasing'.[33]

Thomson, whose advice on the use of colour in gardening 'raised the whole subject to a higher status of art and enjoyment', recommended the use of a decorator's colour wheel. This now common device allowed for

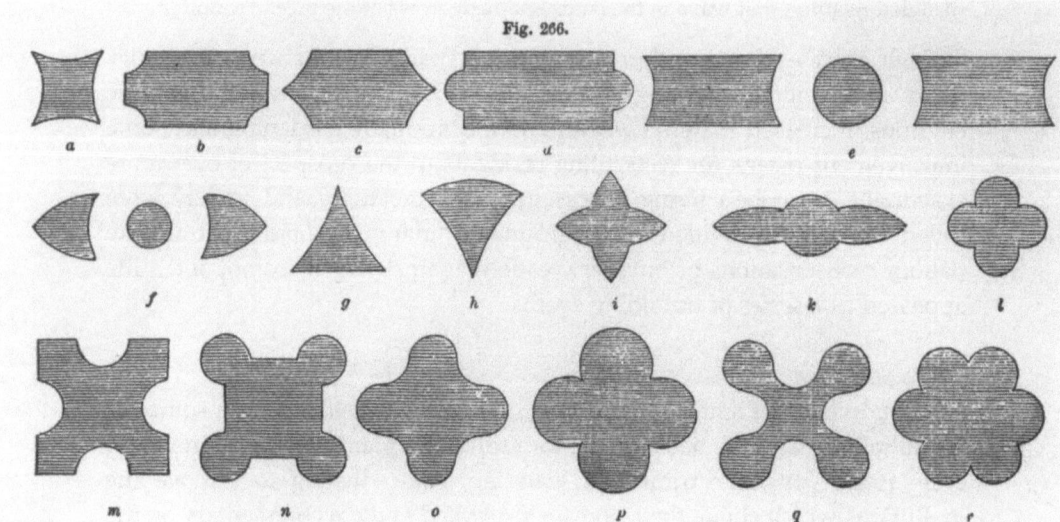

Fig. 266.

5.5 Various shapes for flower beds, in Robert Thompson, *The Gardener's Assistant* (1881)

compromises to be more easily reached between contrasting and harmonizing hues.[34] The domestic gardener could not rely on abstract science or such devices alone, however, to determine the correct application of compositional rules to plants. Flowers were rarely, if ever, entirely monochromatic, but often exhibited a range of hues, requiring careful consideration when choosing complementary species and varieties. These choices varied depending on the setting and the mood the gardener wished to create. A sensitivity to colour placed renewed emphasis on the observer of landscape features as the active interpreter of the scenic characteristics of building sites. Rejecting the 'promiscuous' or mixed style of gardening, Thomson advised the homeowner that:

> Principles of contrasting and harmonising colours must be modified given the particulars of sites. For small beds or groups of beds in quiet sequestered spots where the design can be studied intently, strong contrasts will appear gaudy and should be avoided. Where the garden extends over a wide expanse or is to be studied from a distance, or where there are long stretches of borders, 'every line and figure should live and sparkle with expression, or it will not be easy to follow out and appreciate the length and breadth of the design, as if it were a small picture under the eye.[35]

Prior to this advice, David Hay had asserted that the direct contrast of opposing hues was harsh and unpleasant and needed a 'harmonising colour to mark the full concorde [sic]'. Hay elaborated his design principles in his *Laws of Harmonious Colouring* … (1828) in which he called for a unity of visual experience in gardens and domestic chambers based on chromatic contrasts between plant species and interior furnishings and their surroundings.[36] This practice identified objects of attraction to be seen or read against less obvious surroundings. The effect of such a reading was not dissimilar to what was believed to be the benefit of reading poetry in the Victorian classroom or hearing music from the bandstand of the public park. It was the acquisition of sensibilities and a respect for order appropriate to civil society. Were the science of colour properly cultivated in the domestic garden like music in the park, Hay asserted, 'The turbid and excited mind may be soothed, and the most benign feelings of our nature excited – men may be roused from a state of apathy to attempt deeds of daring valour, or withdrawn from sinfulness to remorse and devotion.'[37]

The subdivision of landscapes into planting beds and the arrangement of plant species, according to chromatic harmonies or other visual criteria, occurred against a sea of grass comprising lawns, terraces and parterres. Grass formed the living fabric of Robert Kerr's 'carpet of design' surrounding his 'Gentleman's House'. It formed a surface on which the garden's compositional organization was coupled with a more complex spatial narrative of paths and clearings and the activities that were imagined to occur there.[38] In *How to Lay Out a Small Garden* (1850), Edward Kemp believed that 'breadth of lawn' was necessary before any notion of a property's extensiveness could be conveyed. As broad a glade of grass as possible should stretch from the best windows of the house to within a short distance of the boundary with as little interruption as possible. Regarding the risk of monotony attending this arrangement, the author of the popular treatise warned that:

Because a very small space, such as a room, will appear larger for being nearly or quite empty, it must not be assumed that a garden is to be judged of similarly; on the contrary, a simple area, which is taken in by the eye at one glance, invites attention to the sharpness of its boundaries. That which requires no mental effort to understand and embrace will never seem extensive, unless of gigantic proportions. The notion of size is not to be realised, within straitened limits, by mere simplicity. It is indefiniteness alone – giving the eye a number of points to rest upon, and recesses to explore, and the imagination a field for its active exercise – that can produce the required result.[39]

Variety was desirable in such spaces and obtained by using serpentine walks of various curvatures where changes of line were hidden from view. Monotony was to be avoided by using single plants and groups of plants and by playing one outline against another, by glades, vistas and recesses and by mixing all sorts of species together.[40] Like a skilful weaver cautioning his apprentice, Kemp reminds the home gardener to pay particular attention to instances where the lawn meets the sides of walks. There it should always be perfectly flat and rise to join a bank or elevated bed as a smooth, concave surface. Where steeper terrain is negotiated and precipitous banks brought close to a walk, rocks or stones, clothed with trailing plants or masses of ivy, should embroider the edge rather than grass.[41] If this were not achieved, a jagged, overhanging border of sward or 'deep harsh line of bare earth' where the gardener has pared it back would reveal the completeness and contiguity in the garden's design to be a sham. Similarly, believing that 'an object of one colour, and that a green one, acquires a striking apparent augmentation of size', Kemp advised that, 'if the plants that flank an open lawn are principally evergreens, and their branches sweep the grass, without any soil being visible, the space is thereby very much increased in appearance'.[42]

STIMULATING ENVIRONS

In the house, like the garden, colour was important in organizing furnishings, wall and floor coverings and ornaments. This was so not only as regards overall interior arrangements, but in terms of relations between interiors and the outdoors. Hay found it odd that in Britain tradition and fashion dictated that white, neutral hues and pale tints of colour were used in interior decorations despite weather that afforded few opportunities for the 'enjoyment of nature's colouring'. His point was that colour in interior design was little used to compensate the British householder for deficiencies of their geography and climate. To the contrary, for the ancient Romans, evidence from excavations at Herculaneum and Pompeii and accounts given by artists and amateurs of the dwelling spaces and remains of interior decoration found there 'all concur in eulogising the scientific knowledge which their colouring displays'. Hence, the ancient Roman's use of deep colours, even black, on walls of rooms lit from above served to offset the brilliant light of the Italian sky. In Britain, on the contrary, Hay charged, the 'vapid tameness in the colouring of our dwellings is the more inexcusable, when we reflect, that as harmonious music delights and refines the mind through the ear, so does harmonious colouring act as an agent of civilization, in delighting and refining the mind through the visual organ'.[43]

Hay echoed Thomson's advice that gardeners should adapt colouring principles according to the mood they wished to create. For support he cited the researches of Isaac Newton (1642–1727), David Brewster (1781–1868) and particularly Johann von Goethe (1749–1832). Goethe had argued that 'Experience teaches us that particular colours excite particular states of feeling.' Accordingly, Hay identified principles of chromatic combination appropriate for each room of the house so that:

In the drawing-room, vivacity, gaiety, and light cheerfulness, should characterise the colouring. This is produced by the introduction of tints of brilliant colours, with a considerable degree of contrast and gilding; but the brightest colours and strongest contrasts should be upon the furniture, the effect of which will derive additional value and brilliancy from the walls being kept in due subordination, although, at the same time partaking of the general liveliness.[44]

Attentiveness to chromatic accents in the arrangement of furniture was but one means for reinforcing the overall use of domestic fixtures to coordinate multiple points of interest and to orchestrate activity within rooms. The result was interior spaces notable for the choices of design and use they offered occupants. These possibilities stand in stark contrast to the more staid life in the 'circle', the highly formalized arrangement of furniture that had governed conversation in the home since at least the seventeenth century. A particularly telling example of the changes occurring in domestic architecture was the demise of this style of interior furnishing. It was an event noted in Humphry Repton's influential text *Fragments on the Theory of Landscape Gardening* (1816). In this book one finds not only attention given to the enlargement and detailing of exterior fenestrations and the visual connections between rooms and the outdoors, but a new way of thinking about and inhabiting space. It was one where it became possible to express a preference for certain activities or forms of contact or for distinctive lifestyles by arranging and occupying rooms differently and more freely. It was where, as Mark Girouard has observed:

Everyday social life was no longer a kind of round game, in which everyone joined in together. Different people could now do different things at the same time and even in the same room. They could drift together and separate, form groups and break them up, in an easy informal way.[45]

Books on domestic architecture and interior decoration, gardening and domestic economy of the Victorian period repeatedly stress the desirability of accommodating novelty, variety and a range of activities in the home. These were first and foremost means for designing stimulating environs all the while maintaining the appearance of order. Certain techniques, spatial motifs or themes are repeated, such as the attention given to one or the other bedding system or mode of arranging furniture. Certain values are emphasized, such as the importance of colour and contrast or the necessity for grouping objects or decorative surfaces to form legible patterns. The point in bringing together the preceding examples is not simply to assert that flower beds were islands in a narrow, metaphorical sense. Neither is it to suggest that Victorians read works by Ballantyne or Stevenson and then consciously strove to equip their gardens with versions of Edenic islands. Nor is it that their predecessors came together in Repton's 'modern' drawing-room like cannibals drawn around a cooking pot – characters brought together by some innate need for community and domestic solidarity. Rather, the point is that both fictional works and domestic treatises were means for developing and sharing distinct modes of self-expression. Ways of interpreting and creating patterns of colour, surfaces and forms brought both imaginary and real landscapes, gardens and interiors together to create a unified and evocative spatial experience.

PERVERTING NATURE

The wealthy found it easier to create variety and numerous points of interest on their estates from combinations of plant species, multiple bedding systems or paths where views and other experiences might be anticipated. These were commonly difficult to emulate in smaller residences in cities and their suburbs. Just as the imprecise boundaries between earth, air and water in the home provoked uncertainty as to the exact source of harmful agents in the domestic sphere, attempts to arrange species and inanimate objects in a coherent way within limited bounds were accompanied by further troubles. There was particular ambiguity regarding the question of just where nature ended and the built environment began. The use of sealed glass cases invented by Nathaniel Ward to safely transport plant species on long voyages did much to promote the use of exotics and has been cited for reproducing the kind of closed world found in glasshouses, but on a much smaller scale.[46] In the relative isolation of these it was possible to observe and imitate the 'marvellous counterbalancing actions of animal and vegetable life that are ever going on in Nature, both in the air and the water'.[47] Easily overlooked, however, were their inventor's grander intentions that they be enlarged to house members of the human species – particularly in their urban habitats. These were intended to serve as popular sanatoria or 'a sort of artificial Madeira, or any other climate'. Placed in windows, the 'Wardian case' was to alleviate the apparent oppressiveness of lower-class dwellings by forming a screen of verdancy, through which views of London's slums would appear as distant ruins, rather than persist as reminders of ruined lives (see Figure 5.6).[48] While Ward's vision failed to materialize, his goal of a verdant screen for every home was carried out on a smaller scale in numerous window boxes that prefigured the rise in popularity of terrariums later in the century.[49]

Pet keeping afforded opportunities to construct artificial climates of a different sort, to transform specific enclosures into habitats and household vermin into species worthy of human care and attention. Like Doctor Moreau's island, the home could be both a cage and an observatory in which lesser creatures were made objects of human values and desires and subject to human anxieties (see Figure 5.7). They provided models of good housekeeping and domestic enterprise. Hence, the mouse, capable of mutilating the Stilton at table or of tunnelling through the loaf of bread, did so 'as neatly as though they had studied under Sir Isambard Kingdom Brunel'.[50] For the urban resident who rarely, if ever, saw the country and had few opportunities, according to one author, 'of witnessing the interesting operations of nature' close at hand, the 'aqua-vivarium' or aquarium was not only edifying, but entertaining. It was neither expensive nor difficult to maintain – so advises *Cassell's Household Guide*:

As the study of all animal and vegetable life presents to the mind a special and elevating influence in addition to the interest it excites, it is a subject of personal gratitude, that the principles upon which the structure of the aquarium is founded have been so carried out and simplified, that this little world in miniature may be adapted to any scale, and that in place of the bowl in which gold-fish were formerly imprisoned and doomed to a slow consumptive death, we can adorn the parlour window with a self-supporting lake, in which the denizens of the water imbibe their natural food, and breathe the gases necessary to their healthy existence.[51]

Careful observation was not only part of learning from life in the fish tank; attentiveness was crucial, so that, among other requirements, the aquarium should present an orderly appearance. Muddy or cloudy water was not allowed, since it could make it as difficult to distinguish 'one finny inhabitant from another, as it is to recognize a friend in a London November fog'.[52] Success in this department served, in turn, to reflect well upon the general character of the pet owner, the gardener, terrarium or fern enthusiast. It revealed their 'thrifty and industrious habits' or even their worth as a member of society:

Those who begin by loving Nature and inquiring into the structure, organization, and habits of plants and animals, and the wonders of the inorganic kingdom, will

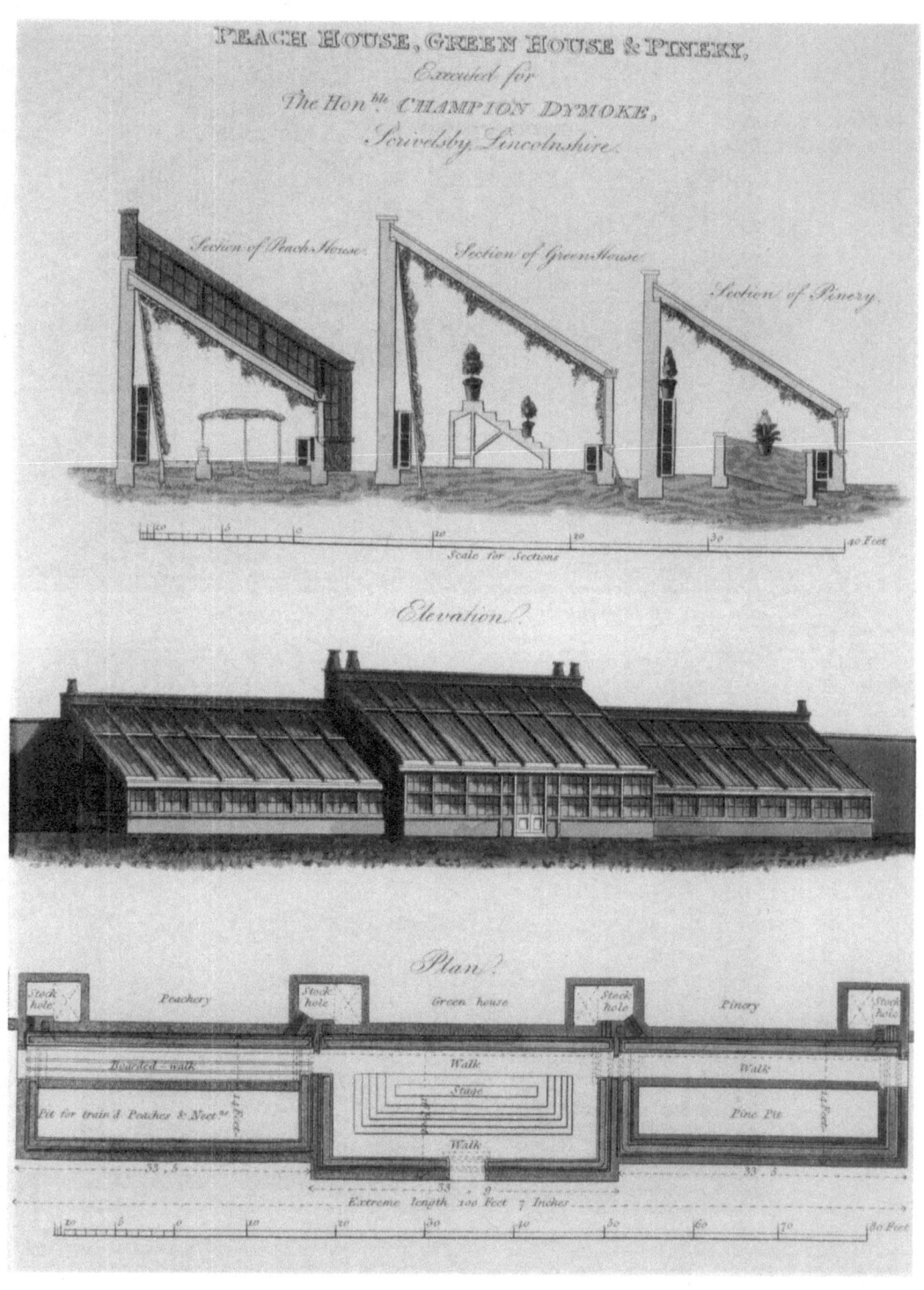

I   Sections, elevation and plan of a peach house, a greenhouse and a pinery, in George Tod, *Plans and Sections of Hothouses, Green-Houses, An Aquarium, Conservatories, &c.* (1812)

II Joseph Wright of Derby, *Experiment with the Air Pump* (1767), oil on canvas, 183 × 244 cm (72 × 96 in)

III The Palm House, Royal Botanic Gardens at Kew, by Decimus Burton and Richard Turner, 1844–48

IV Joseph Paxton, perspective drawing of a section of the 'Great Victorian Way', 1855

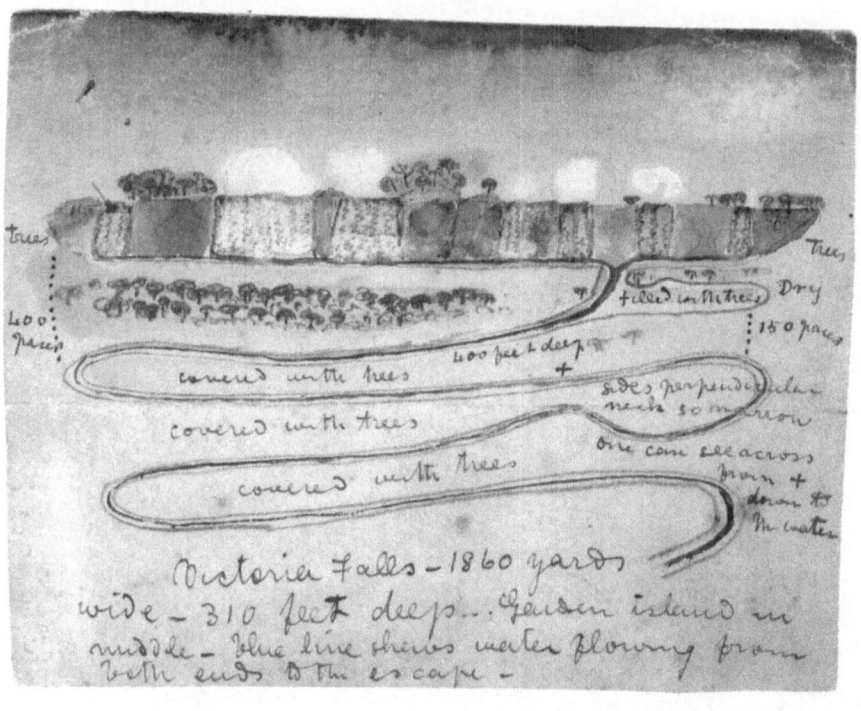

V David Livingstone, sketch of the Great Cataracts of the Zambesi River known as Victoria Falls, 1855, watercolour. There is a sense in this watercolour of its author attempting to come to grips with an immense and unfathomable landscape where, as related in his journal, viewed from 10 km away from the falls, its vapour or 'smoke' seemed to mingle with the clouds

VI The Eden
Project, near St
Austell,
Cornwall

5.6   A fern-case designed as a window of Tintern Abbey, in Nathanial Ward, *On the Growth of Plants in Closely Glazed Cases* (1852)

soon be animated by kindlier feelings towards all around them … Let parents try to inspire their children – the best of all 'Home Pets' – with a fondness for natural science; whether it be encouraged by keeping and caring for a dog, a cat, a rabbit, a pigeon, or a song-bird; by rearing flowers; by forming an herbarium, or a collection of moths and butterflies, or by other kindred means, and they will surely be better boys and girls, and make better men and women, better members of society, and above all better Christians.[53]

5.7 'Pets of the Aquarium', in George Kearley, *Links in the Chain; or Popular Chapters on the Curiosities of Animal Life* (1862)

Useful instruction and imaginative play were combined in these miniature worlds. They were one means of acting out a virtuous life by equipping the homes and sustaining the lives of lesser creatures. Wardian cases and window boxes, terrariums and aquariums – like the house and garden more generally – were vehicles for reproducing the relationship of humanity to the organic world on a small scale. By caring for these lesser places, homeowners and their children would learn to care for their own surroundings. The challenge of equipping window boxes, glass bowls and wire cages with the requisites of life, like the task of improving poor soils or bringing light to small yards, tested the ingenuity and perseverance of the resident as well as the domestic gardener. For both, there was little excuse why 'a foot of soil should lie idle' or, in the home, any space be wasted.[54]

In attempting to reproduce the ordered arrangement of nature in the home – to contend with the immensity of the organic world and the limited scale of the domestic one – householders were made ever mindful of artifice. Kemp warned the small homeowner that:

To regard a garden otherwise than as a work of art would tend to a radical perversion of its nature. It is and must remain that which its proximity to the house alone enables it to be. No ingenuity can convert it into a forest glade or a glen. Nor is such transformation to be wished for, were it possible any more than that a dwelling should be transmuted into a hut, or a den, or a cave.[55]

Even the common aquarium could reveal a world of deceit and pretence, for just as 'we neither find cowries, corals, nor busts of Napoleon in our brooks and ponds, we have no right to introduce such objects in our tanks'. The accurate imitation of habitat demanded that the homeowner should forego such practices as incorporating a hollow within a glass bowl so that goldfinches or linnets might be seen hopping in the midst of water with fish swimming around them, like shark around prey. This was a fashion denounced as 'barbarity'. The exhibition of fish in such a way, though 'agreeable and pleasant' for some, was hardly edifying for being so obviously artificial. Rather, it was 'whimsical and unnatural'.[56]

By the 1860s evergreens, ivy, trees and shrubs such as box, holly and yew were commonly used to relieve the starkness of winter months. This was not a new technique, though formerly only the wealthy on their estates were the main beneficiaries of such designs. By the second half of the nineteenth century, the spread of gardens amongst the middle classes, the widespread availability of a wide range of relatively low-cost, nursery-grown species and cheaply produced manuals on how to choose, plant and care for them turned a larger part of Britain outside the glasshouse green in winter as compared to earlier times. Coupled with techniques for controlling the atmosphere indoors, warming and ventilating rooms and buildings, such horticultural techniques contributed to desires for constancy in their surroundings. For many, the illusion of an endless spring was often completed by use of coloured earths and gravels, and it was popular for a while to form bedding patterns of crushed waste kiln bricks, sifted cinder ashes and sand. Not everyone responded with enthusiasm to the idea of an endless spring.[57] Fear of artifice in the garden led many horticultural experts to criticize such measures though with very different terms to those that might have condemned the alchemist for forcing trees to bear fruit out of season. In some instances, criticism appealed to human psychology. Thomson argued that the desire to present the garden in flower all year led to a 'sameness' which numbed the mind to the 'moral influence' of flowering plants:

> To the attentive eye, each change in the seasons brings its own peculiar beauty and charm. If, instead of change, we had one continuity of song, leafy woodland, and flower garden, would it not become monotonous, and cease to be a source of exquisite pleasure to the mind? Lovers of flowers ought to be thankful that the year and the human heart have room for changes.[58]

In his call for the use of 'harmonious colour' David Hay bemoaned the fact that 'fashion, rather than scientific knowledge' often rules the day.[59] This criticism was supported by theories later in the century that questioned the desirability of highly contrasting colours, calling instead for the use of subtle gradations and neutral shades.[60] These were the same hues that Hay had himself criticized for failing to enliven the dull British sky. A concern for the integrity of places, their variety *and* wholeness, provided a context for arguing for the appropriateness of one or the other form, surface, texture or colour. Accordingly, living with one's environment by assuming its completeness was akin to maintaining a sense of one's own. Living 'honestly' with nature in such a way was a means of asking questions, not necessarily a source of answers.

Habitats and ways of life

Attentiveness to nature's fundamental elements in the home was coupled with respect for the complexity of the environment, its unified character and the suitability of each locality for the living beings found there. Environmental awareness furthered expectations that domestic surroundings be designed with completeness in mind as well as a distinctiveness based on the individual needs of residents. Such sensitivities were encouraged by works of science and fiction as well as treatises on domestic architecture, gardening and household economy. Few of these works used the term habitat to describe the unique relationship of living beings to their surroundings or characters to fictional settings – fewer still to describe the dependence of residents upon their homes. Nonetheless, these related ideas were implied in much nineteenth-century literature. They came to be expressed through the careful arrangement of plant specimens in the garden and furnishings indoors, the garden flower-bed and the common aquarium. An understanding of habitat, however partial or intuitive, informed a particular way of reading domestic surroundings. It informed a manner of interpreting enclosed or walled spaces, the closeness or distance of objects and possessions in terms of how one might live amongst them.

The desirability of variety in the Victorian house and garden and its role in accentuating a coherent and distinctive spatial experience cast the domestic realm as one corner of a broad field for imaginative exploration and self-expression. Along with the expectation that the home should provide a

protective retreat for the body from the rigours of the street and the
workplace was the demand that it stimulate or 'enlarge the mind'. Along
with it being a 'lesser breathing space' to ameliorate the excesses of
urbanization was its role as a 'special and elevating influence' – a place where
significant events happened in life. In *Our House and Garden* the author
equated an ever-closer scrutiny of the home with travel, as the unknown was
something to be experienced both nearby and far away, for

we may probably find, even around our own firesides, deeply interesting
phenomena, of whose existence we had no suspicion. We may discover currents as
interesting as the trade winds, or the gulf stream, or other ocean phenomena. We
may notice effects of different degrees of temperature, as curious as those witnessed
in the arctic, or the equatorial seas – and even in animal and vegetable life, hidden
mysteries – beautiful and beneficent arrangement, which might well enlarge our
minds, and show us how to improve the tone of our physical and mental powers,
without seeking for these results by distant excursions.[61]

Johnson's encouragement to observe nature at home bound the house and
garden into a contiguous whole where every part was seen to relate to
another. His admonishment to 'but think a little' about the operations of the
organic world fixed the householder and homemaker as the home's most
important inhabitants at the same time. His was a demand that residents
respond appropriately with both creativity and restraint in designing a home
for themselves amidst animate and inanimate matter.

Though ever more consciously bound by walls and routines of care, the
environment exceeded the confines of the Victorian house and garden, just as
it had the glasshouse. Attention was directed alternately both outwards and
inwards. It was cast between landscapes close to hand and ever larger
domains with the realization that the well-being of residents frequently
depended on factors beyond their immediate experience and control. The
idea that the home should 'blend insensibly' into nature or that domestic
gardens appear limitless was only possible with endless labour and artificial
means. Achieving both variety and wholeness commonly involved the
complete reshaping of existing terrain, the importation of plant species from
abroad and the decoration of interiors to enliven the dull climate of Britain.
Adapting to the circumstances of particular sites often entailed appropriating
features of other places altogether and the ways of living found there. One
might imitate the Romans, for instance, in their skilful and 'scientific' use of
colour, if not their use of certain hues per se. Questions of the artifice or of the
'unnaturalness' of such appropriations inevitably arose. The idea of living
with nature by standing apart from it and observing its laws more clearly was
impossible to fully implement. No amount of soil enrichment, smoke
abatement or water improvement, planting or weeding, painting or
decorating could negate the fact that for most residents of Britain's cities their
world was a densely populated and increasingly globalized one. It was a
world where many of the requisites for life came from somewhere else.

The limits of environmental mastery and bounds of artifice were raised by
William Robinson in his best-selling manual, *Hardy Flowers* (1871). There he

advised the use of indigenous plant species as an alternative to extensively manicured, bedded and costly managed grounds:

It is to me a cause of surprise that while we find persons going to great expense to build a glass box wherein to preserve a little of the pretty vegetation of New Holland and other warm climates, and which is of necessity always in a condition less beautiful and less satisfactory than vegetation flourishing in the free air ... We live in a country which is, on the whole, better calculated for the successful culture of the most beautiful vegetation of northern and temperate climes than any on the face of the earth, and at present we take as much advantage of it as if we lived in one where, from extremes of some sort, such vegetation could not exist, and where extraordinary and expensive artificial means were requisite for the enjoyment of a little vegetable beauty.[62]

Robinson bemoaned the considerable expense and labour required to mimic the tropics and other exotic environs. This was to ignore the economy of nature where every species has its place. Drawing on the lessons of the natural sciences, botany and zoology, his advice to horticulturalists was easily directed to architects and their clients. Seemingly unable to thrive in the open air like Robinson's hardy flowers, the British were threatened by moral degeneration in their own, poorly designed homes. It was a threat John Ruskin associated with the slapdash state of much of the architecture in his day.

In 1837, in one of a series of essays written for Loudon's *Architecture Magazine* (1834–38), John Ruskin described the proper relationship between human beings and their surroundings. It was a wholeness of form, function and expression that was reducible to no single rule. This entity united an 'all unity of feeling' unique to humankind, on the one hand, and a 'poetry' which was its architecture, on the other.[63] Without this overlap of living and social being, he bemoaned the consequence of seeing 'nothing but incongruous combination' everywhere. Ruskin did not use the term 'habitat' to describe this dependence of human beings on their surroundings. It is worth noting, however, that his thoughts on issues like taste and architectural beauty were amenable to subsequent appeals to the connectedness of form and locale. It was an idea reducible neither to mere shelter nor expressive impulses, buildings or gardens alone. Rather it encompassed places just as suitable to each race or nation as every geographical area, station or spot of ground was to the species that lived there. Ruskin's thoughts on this and related subjects were subsequently developed in his book, *The Poetry of Architecture*, published in 1892.[64] There one learns, amongst other aberrations of human and organic nature, of the author's abhorrence of the bedding out of flowering plants:

A flower-garden is an ugly thing, even when best managed: it is an assembly of unfortunate beings, pampered and bloated above their natural size, stewed and heated into diseased growth; corrupted by evil communication into speckled and inharmonious colours; torn from the soil which they loved, and of which they were the spirit and the glory, to glare away their term of tormented life among the mixed and incongruous essences of each other, in earth that they know not, and in air that is poison to them.[65]

Recall that a comparable feeling for the suitability of living things for their locale led the curator of Kew's newly completed Palm House to criticize its incongruous combination of iron rafters, girders, galleries and pillars. He likened the structure's appearance to a smithy or railway station, rather than the proper abode for plants. Similarly, Ruskin, in the first year of Queen Victoria's reign, was led to condemn the state of Britain's built environment, where he found

pinnacles without height, windows without light, columns with nothing to sustain, and buttresses with nothing to support. We have parish paupers smoking their pipes and drinking their beer under Gothic arches and sculptured niches; and quiet old English gentlemen reclining on crocodile stools and peeping out of windows of Swiss chalets.[66]

The natural theologian, the botanist and evolutionist as well as the gardener and architectural critic were each fascinated by the diversity of creation. They were concerned with the position of human beings in the natural world. Use of the term habitat is found in works of eighteenth-century natural history, where it generally denotes the locality in which a plant or animal grows. By the middle of the nineteenth century it had acquired the additional meaning of dwelling place. The American poet James Russell Lowell wrote in 1854 that 'every thing is not a Thing, and all *things* are good for nothing out of their natural *habitat*'.[67]

The idea of habitat and thoughts of the appropriateness of the ways of life the British had acquired over their long history for unique places like Britain itself were strengthened by sensitivities to the environment. These became pronounced in the nineteenth century, particularly during the Victorian era, when they joined both biological and social spheres of being. Understanding the economy of relations between living things and inanimate matter, human beings and lesser species forming the organic world, however, privileged no single interpretation of what the built environment should look like – apart from a general expectation that it appear whole, contiguous, ordered and, ultimately, interesting. The home *was* an interesting place because it *could* assume a variety of forms, all the while serving its primary purpose as an orderly setting for domestic life. 'Newness' was very much a part of that life. It provoked one to discriminate between the spontaneous arrangements of organic nature and the thoughtful activities of humankind. Newness was found in differences between the number and forms of species dispersed across the surface of the earth and the varied ways humans adapted to their own place on the globe. It was experienced over time, cast between the movements of islands and oceans, rivers and mountains and the ceaseless change of everyday life.

Characterizing life at home

The appeal to biology and the natural sciences, physics and chemistry, landmark texts and everyday, practical manuals made the Victorian householder and homemaker a scientist of sorts, familiar with nature's elements and forces in some degree. With works of erudite theory and fiction by their side and sensitive to the variety of living forms as well as to the apparent wholeness of the natural world, residents were encouraged to design gardens and furnish rooms in like manner. Reading about unique places and wanting to create them for themselves, homeowners found that they could develop their imagination and creative talents, all the while providing for family security. Of course, confidence founded on the mastery of the home was undermined by fears that domestic life was not so easily controlled.

Environmental awareness was not only a matter of being aware of organic vitality and the animate and inanimate things lying about the house and garden. It was also a matter of consciously positioning oneself within environs of varying scales to maximize comforts and minimize anxiety. This could happen in a number of ways, as previous instances illustrate. Actions aimed at improving garden soils or removing air-borne dust, affording ventilation or forecasting the weather, becoming vegetarian or turning parks into breathing spaces in the city addressed local as well as global concerns. One set of issues often overlapped the other. Uncertainties regarding the imprecise boundary between natural and built environments were really questions about just where one stood to derive the greatest benefit from one or the other.

More often than not, observations of the varying contexts for organic or physical operations were guided by mental efforts to predict, enhance or forestall the consequences of actions and choices. If one only knew *which* house to buy, for example, then unhealthy exhalations need not arise, cracks need not appear and draughts need never enter the home. This chapter explores how life in the Victorian home obliged residents to question likely sources of dissatisfaction in their daily lives and then deal with them. It considers one, easily overlooked way that these questions were raised, if not fully answered. This was the imagining of characters by authors and their readers, designers and their clients, to inhabit their plans for the ideal house and garden. This chapter focuses on one book where this practice is clearly advocated, Robert Kerr's *The Gentleman's House*, first published in 1864.

Addressing Victorian homebuyers and builders, Kerr suggested that no room could prove satisfactory until its designer had 'in imagination occupied it and proved it comfortable'. The author, known more for his academic pursuits and architectural advice than for any realized building, invited readers to explore his carefully delineated rooms and garden terraces in their minds. This was an important first step before affirming the suitability of rooms for ease of living or before even thinking about their style or manner of decoration. Relying, in effect, on a kind of literary characterization or deceit, Kerr called upon the figure of the retiring gentleman to detail principles of good planning. By following the movement of this figure from bedchamber to parlour to parterre, one pictured the arrangement of rooms and corridors, house and garden, that best made the home a well-formed whole, a source of comfort and pleasure. Kerr's seemingly straightforward advice, familiar to architects and landscape architects today, belies a more profound assertion regarding one aspect of human perception and experience, namely, that both were linked and intrinsically 'spatial'.

The settings of fictional narratives were useful in bridging the divide between the idea of a place – be it verdant like an island Eden, inviting like a river valley or hilly in appearance – and a real place created in the house and garden. The concept of habitat provided a context for both imagining places and building them, for questioning the limits of human interference in the organic world and the suitability of the built environment. Evidence for the existence of habitats and ways of living arising in them brought together the concerns of ecologist and designer alike. For the former, a locality was known by the uniqueness of its animals and plants. For the designer, a residence housed the unique lifestyles of its occupants. In this chapter literary characterization is discussed as an important bridge between a sense of a space and thoughts on just how one might actually live in it. Even if all the physical requisites for a comfortable domestic life could be met in the home, there remained other decisions to make. Questions invariably arose as how to live virtuously, for instance, with an economy of means or in a typically 'English' manner.

Much has been written about how the Victorian home organized public and private life, family relations and social conventions. In this regard *The Gentleman's House* is encyclopaedic in detailing the structure of domestic life, particularly among Britain's more prosperous classes.[1] Such works on architecture, like those on gardening and domestic economy, supported a growing belief that human beings were burdened with a unique need to locate themselves in the world, spatially and temporally. Long before psychologists, behaviourists or sociologists developed theories to explain the role of perception in such terms, it was known by such commonplace descriptions as the 'endless confusion' caused by houses filled with dust and grime. It was evoked by the claims of the murkiness of a London fog or uncared-for aquarium or, for Philip Gosse, even the 'inextricable confusion and profuse luxuriance' of the glasshouse that made it into 'a great whole' – a place of grandeur as well as potential disorientation.[2] As it was cultivated in the home, spatial acuity was a matter of moving and looking through passages, doors and windows to rooms and exterior landscapes beyond. It was also a matter of looking inward. Through both scientific and literary means, perception acquired a more thoroughly psychological cast, providing a means for recognizing the effect of spaces on feelings and sentiments.

Visualizing how they might live in a home, assisted by practical manuals, plans and abstract goals like achieving 'comfort' or 'compactness' in designs, architects and landscape gardeners as well as ordinary Victorians were able to anticipate their reactions to houses and gardens. Establishing principles of good planning in such a way, design was itself made a more objective and, it was hoped, certain practice. Characterizing the inhabitant of domestic space in such conceptual or abstract terms means that he or she was related to their surroundings, not simply as a plant to its glasshouse or goldfish to its bowl. They were not simply creatures that reacted passively to the environment in a purely physiological sense. Rather, a resident's position within their environment was determined by forms of self-awareness, memory and discernment. Accordingly such notions as convenience and comfort, like the value and pursuit of health and well-being in a biological or medical context, necessitated choice and the exercise of free will.[3]

While one can imagine some dim-witted architect of the 1860s failing Kerr's test of mental occupation, it is hard to imagine today a designer who configures rooms and garden landscapes without some consideration of how one moves from one to another. It is hard to neglect some measure of convenience, privacy or accessibility. These qualities have long since established the home as a place uniquely suited for human occupation.

A gentleman's house

Of the countless books written on the subjects of domestic architecture and gardening, Robert Kerr's *The Gentleman's House* is worth considering. It is notable not only for its weightiness, but also for the author's strong advocacy and methodical study of good planning and design.[4] Of course, thoughts on good design were neither new to the Victorian era nor fully accounted for in Kerr's text. Treatises written in earlier times also urged the careful disposition of rooms – encouraging their 'convenience', as it were – particularly those entailing plans for rural cottages and workers' housing. The great proponents of nineteenth-century design – figures such as Augustus Pugin, William Morris and Ebenezer Howard – are commonly cited by historians. The influence of the authors of pattern books, such as Philip Webb and Charles Voysey, is equally well known. Kerr's book is worth considering more closely, though. Along with manuals by similar, less-celebrated writers, it was particularly aimed at Britain's broad house-owning classes. They were the residents most likely to benefit from improvements in the domestic sphere. Kerr's manner of argument, moreover, is worth examining in some detail. It draws upon a series of historic or well-known residences, carefully delineated and accompanied by extensive commentary detailing the qualities of their spaces and manner of arrangement. Counterpoising these, more practical and 'scientific' plans are detailed. As well as interrogating history in such a way, the author urges designers to consider the interdependence of rooms and exterior spaces. He uses diagrammatic tools and terms such as 'comfort' to do so. While very much a product of its time, the book seems remarkably modern in its concerns.

Many of the historical residences cited in the text were inspired by the sixteenth-century Italian architect, Andrea Palladio, a designer commonly associated with Classical values of simplicity and elegance. Some residences were taken from the treatise of his English admirer, Colen Campbell's *Vitruvius Britannicus*, published in multiple volumes from 1715 and onwards over the course of the eighteenth century. It is perhaps incongruous that Kerr's support for sound planning in architecture over issues of style and historic precedent depended so heavily on aristocratic dwellings for exemplification. However, these were laid bare, usually indicated as simply delineated floor plans unaccompanied by elevation or section or the slightest hint of neo-Classical ornament. Having had their wings pinned like great stick insects on a drawing board, one finds in Kerr's text Sir John Vanbrugh's Blenheim Palace (1715) struggling vainly for prominence alongside William Kent's Holkham Hall (1734). Courtyards, chambers and cavities have been stripped of façades, laid bare for the masses who could have never hoped to experience life inside. They could have imagined, however, with Kerr's help of course, the many discomforts contained within (see Figure 6.1).

The Gentleman's House begins with a history of the evolution of the modern English house and a statement of its three chief attributes: privacy, careful arrangement and what the author called the 'catalogue of rooms'. In describing these, Kerr believed that the first two were little advanced by the eighteenth century. The third, however, had a long history. The 'catalogue' or differentiation of space into halls, galleries and various chambers and parlours was found in the noble dwellings of the sixteenth century, for instance. But then these were 'too indefinitely contrived, as regards their precise use and their relation to each other in disposition'. In the author's view, there was little room in these palaces for the modern notion of convenience. A work such as *Vitruvius Britannicus* included a variety of handsomely illustrated mansions, commonly exhibiting the symmetrical, rectangular subdivision of interior spaces and nobly titled rooms found in the Palladian style. It remained, nonetheless, a directory of the 'waste of space' characteristic of period planning. It proved to be the 'ingenuity' of the nineteenth century that made sense of the forms of accommodation housed within the eighteenth-century manor. Arising from the inchoate plans of

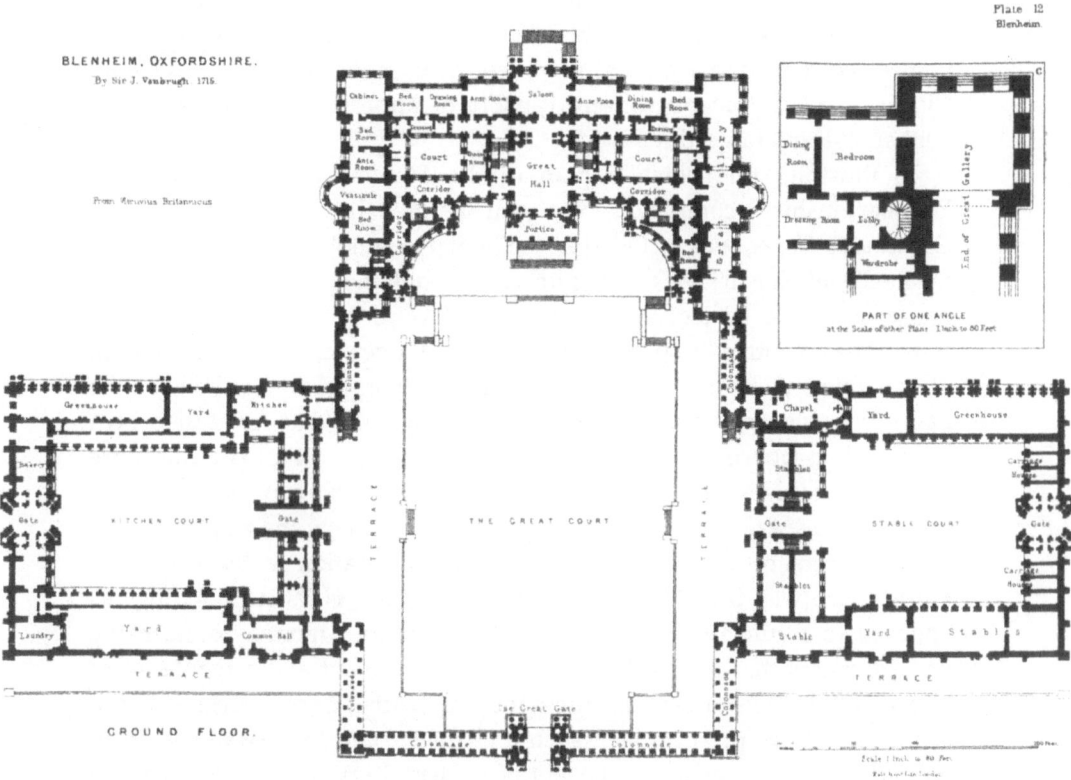

BLENHEIM, OXFORDSHIRE.
By Sir J. Vanbrugh 1715.

From Vitruvius Britannicus

GROUND FLOOR.

Plate 12
Blenheim

PART OF ONE ANGLE
at the Scale of other Plans 1 Inch to 80 Feet

6.1   Sir John Vanbrugh, Blenheim Palace, Oxfordshire, principal floor plan; in Colen Campbell, *Vitruvius Britannicus* (1715)

Campbell and his contemporaries, the succeeding century's 'scientific mode of adjustment and arrangement' was the chief subject of Kerr's book.[5]

Histories of nineteenth-century architecture often cite the growth of historical eclecticism and the so-called 'battle of the styles' that prompted much debate during the Victorian period.[6] For Kerr, the historical revivalism of the first half of the century arose from a general freedom of thought, particularly as this heralded new opportunities for designers. Revivalism was a challenge to the dominance of Palladianism and so, initially, the movement was a good thing. It soon 'ripened', however, into a narrow antagonism between the choice of either neo-Classical or Gothic forms for buildings.[7] This was pointless and misleading in Kerr's view, for such a narrowing of options obscured other principles of sound design. These were not reducible to simple, stylistic concerns.

In Kerr's estimation, the popularity of Classical precedents drawn from ancient Greek and Roman sources had little direct impact on the evolution of sound, nineteenth-century planning. While the Palladian style possessed a certain stateliness, its advocates were prone to be pretentious and not in keeping with true English character. The Gothic, on the other hand, proved amenable to the unique English propensity for comfort and convenience.

Readers of *The Gentleman's House* were told that the revival of the Elizabethan style in particular, in opposition to neo-Palladianism, was largely due to the superiority of its planning, but also 'to no small degree to the desire of the English gentleman to avoid obtrusiveness'.[8] Kerr concluded that while both Classical *and* Gothic plans were capable of affording domestic comfort, neither style guaranteed it as such.

To illustrate this point, a plan of the Manor of West Shandon in Dumbartonshire, completed one year before the publication of *The Gentleman's House*, was included in the book to describe the peculiarities of the medieval revival style (see Figure 6.2). Though the principal rooms of the house are more or less symmetrical, they were 'purposefully irregular, sometimes eccentrically so', according to Kerr. If one were to indicate lines of 'thoroughfare passage' – to trace with a finger along the plan or imagine the paths of movement through the house – they would: 'wander at their own sweet will in labyrinthine freedom quite beyond the reach of art. The entire composition gives one the idea of a rabbit-warren; you can get from anywhere to everywhere at a jump – provided you know the way.'[9]

The difference of value realized by this illustration is the difference between the way a rabbit or another species inhabits its surroundings as opposed to a human being. Unlike other creatures, humans presumably needed spaces that offered both a *choice* of movement and a regularity of planning that allowed them to 'know the way' with a degree of convenience and certainty. By presupposing such a character and then 'walking' them through various plans for the benefit of readers, Kerr was able to diagnose the inadequacies of one design versus another. On this basis, he was able to conclude that both the neo-Classical and Gothic styles could exhibit the qualities of good planning. On the basis of their composition, for ease of movement and economy of rooms, they need not be considered antithetical, for:

to live in the one would be precisely the same as to live in the other; in a word, one might choose between them by lot (at least such is the intention, whether successfully accomplished or not); and yet this is our argument, – that the one exhibits throughout an all-pervading *balance* which need not be constrained, and the other an all-pervading *freedom* which need not be unruly, as two distinct styles of Plan between which there seems to be thus far really no difference of value.[10]

Interestingly, Kerr capitalized the first letter of the terms 'Plan' and 'Thoroughfare' in his book. He used the terms in a singular and indefinite sense to describe a timeless quality of domestic order, little evidenced in earlier times though much desired in his own. It was an order based on certain compositional values manifest in plan drawings and on a particular kind of spatial acuity and sense of orientation. Kerr reinterpreted the now well-trodden clash of architectural styles in terms of each being simply a different way of configuring rooms. In *The Gentleman's House* the neo-Classical and the Gothic appeared as variations on the same theme of suitable habitat. Irrespective of which style clothed the exterior of a building, an organic unity of spatial experience was presumed indoors.

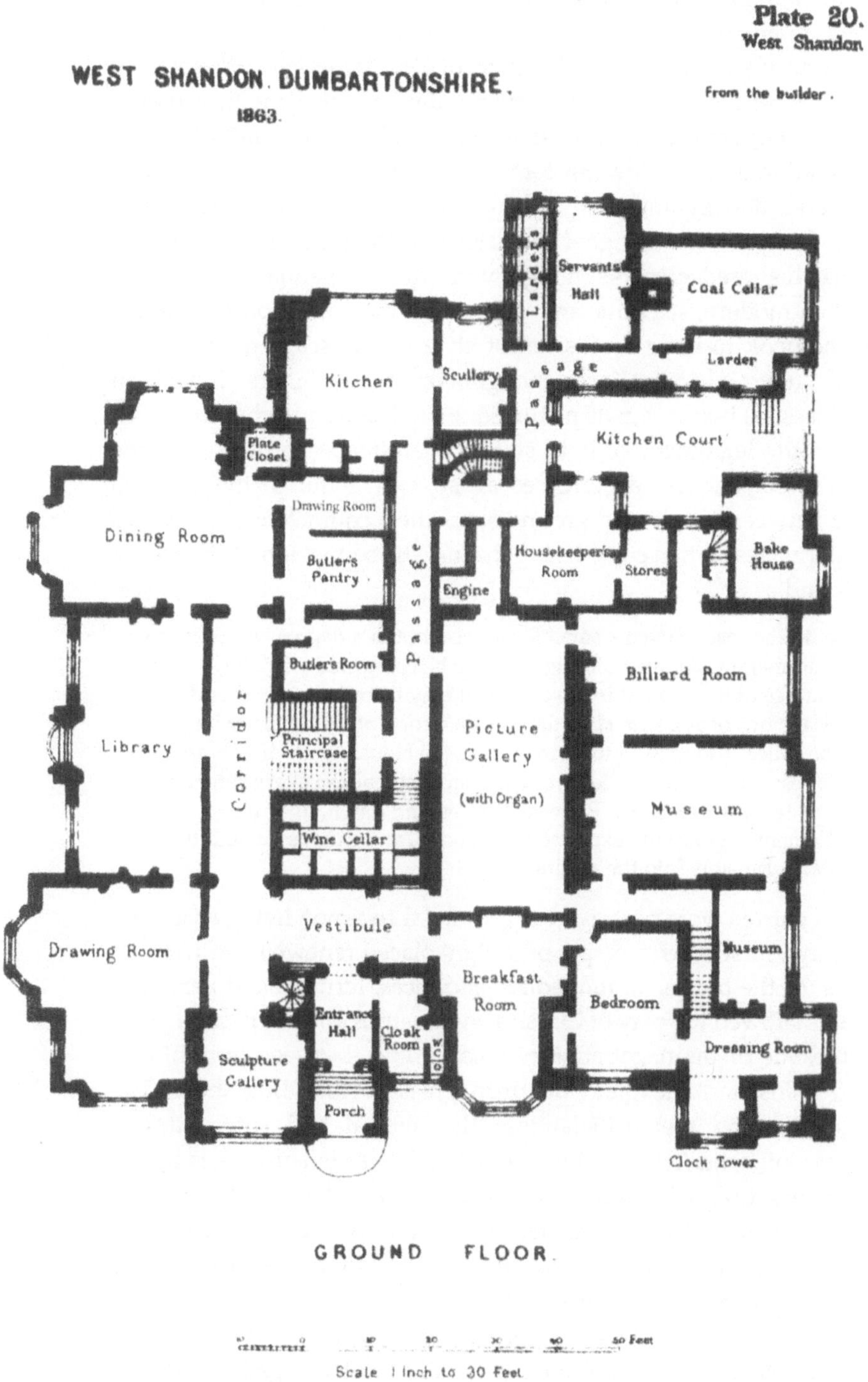

WEST SHANDON. DUMBARTONSHIRE.

1863.

Plate 20.

West Shandon

From the builder.

GROUND FLOOR.

Scale 1 Inch to 30 Feet

6.2 Manor of West Shandon, Dumbartonshire, ground floor plan; in Robert Kerr, *The Gentleman's House* (1871, ₁1864)

ACCOMMODATING LANDSCAPES

This sense of habitable space was reinforced in the book, where Kerr's plans were thrown open to the outdoors. The author's comments on landscape gardening occupy a relatively small section of *The Gentleman's House*, as he began by advising the prospective home-builder to consult with a proper landscape gardener at the earliest possible moment. His comments alluded to the transformation of the eighteenth-century horticulturalist, concerned with regular surfaces and symmetrical avenues, gardens and terraces, into a professional designer. The designer 'luxuriates in the play of nature's own features' like his predecessors, though with Kerr's manual in hand he proceeded in a modern, scientific and methodical manner. A portion of this section of the book forms a discussion of style in landscape gardening. It parallels, in condensed form, Kerr's thoughts on architectural style with an analysis of Classical Italian versus picturesque English approaches to garden design. Given the legitimacy of both styles of landscape gardening 'in the hands of an intelligent and experienced artist', each required the practical dependence between house and grounds and their common design.[11] This was particularly clear when considering the 'artistic connexion of the House with the ground':

> To some extent in the case of even a small Residence, but in a degree which increases with its style and magnitude, the building ought to be connected with the surrounding surface of the ground in a way which may be called artistic; and in dealing with Mansions of superior class the utmost efforts of the designer have frequently to be called into request to form around the House, as itself only the central object or casket, a carpet of design, which shall spread on every side in the various forms of Terrace and Court, Parterre, Garden and Lawn, until the architectural element is gradually expanded, expended, and exhausted, and the artificial blended insensibly into the natural.[12]

House and garden were portrayed as forming a common living space in a number of ways. First, Kerr's 'carpet of design' placed renewed emphasis on the 'adjuncts' of the house, its immediate entrances, terraces and parterres. These features allowed the play of Classical symmetries and the natural grace of the picturesque or the negotiation of incongruities between upright and ornamented walls and the plain of green grass surrounding them. The principle of blending the artificial into the natural had some curious repercussions. Soil, when not suitably planted in the garden or formed into 'a bond of combination', exhibited a 'sort of nakedness that cried out to be covered'. To the contrary, under the 'refining control of art', the irregularity of nature was permitted to approach 'almost to the door'.[13] Second, in that compromise and blending are portrayed primarily as a matter of planning the book states that designs must be carried out in considerable detail. This ensures that the fabric of any building was fully knitted together. Flower-beds were to be introduced in any of the 'recesses of plan', though they must be used judiciously. If not, then a terrace might be mistaken for a parterre; that is, the overall legibility of the chosen architectural or landscape style or the function of a particular element of the design might be misconstrued.[14]

Third, and perhaps most importantly, a position for each 'spot of ground', like that of every interior apartment, was sought to maximize its exposure to sunlight despite being arranged with desirable views and privacy in mind. This suggested a distinction in the mode of occupying space, between those that were seen and those that moved through.[15]

By calling to mind the figure of the inhabitant, Kerr's reader and would-be designer were compelled to effect compromises between interiors and exteriors based on an imagined spatial experience. Distant views and immediate exposures became significant. Ease of movement was facilitated, requiring careful consideration of ground levels between rooms and adjacent terraces or courts.[16] Conversely, though communication between rooms and terraces was best facilitated by establishing level ground throughout, stylistic integrity demanded that the latter be visibly distinguished from surrounding lawns. This required a change of elevation and the provision of either a balustrade or grass slope. A *promenade*, that 'intelligible and useful adjunct' to even the smallest of houses, useful in foregrounding its façades, required both an agreeable view and shelter to justify its inclusion in the design. So as not to subject the residents to undue anxiety during their walks, promenades were to be straight, though they might retain a *variety of outline* by the introduction of bays or bastions. If flower-pots, shrubs and statuary, or any other ornaments be used along walks close to the house, these were to be arranged in a regular manner. Their planning should form an integral part of the surface connecting 'the architectural character of House, and perhaps Terrace, and the landscape character of the further Lawn'.[17]

ASPECT AND PROSPECT

To negotiate this geography of inner chamber, border terrace and 'further Lawn' Kerr's reader and prospective designer were provided with the author's 'Aspect-Compass' as an aid to the imagination. In a way the device is similar to John Loudon's diagram of the position of houses relative to the sun in that it directs thoughts to the immediate environs of the residence. However, it made a more direct contribution to developing a sense of the experience of interior spaces. The compass was a means for determining 'scientifically' the most suitable relation between two distinct contexts for experiencing a place. One was the position of walls and windows relative to their exposure to sunshine and weather – this context being their *aspect*. The other concerned the relative position of the corresponding room to the surrounding landscape and the qualities of light in which exterior views were seen – this was called the room's *prospect*. With a schematic plan of a south-facing window at its centre, the aspect-compass charted the bearings of various climatic phenomena. It recorded the hours of sunshine for different exposures, the positions of sunrises and sunsets throughout the seasons, and the directions of winds, whether boisterous, cold or mild (see Figure 6.3). Given the standards and expectations of the day, this meant that ideally a house should have its drawing-room façade turned to the south-east. The dining-room should face the north-west and offices the north-east. The south-

6.3 'Aspect-Compass', in Robert Kerr, *The Gentleman's House* (1871, ₁1864)

west would be left free for any 'compensation required by necessities of plan'.[18]

Despite the desire that the 'artificial blend insensibly into the natural' in the plan, the inhabitant of this terrain was not to be so easily accommodated. As Kerr pointed out, the effect of aspect on a room and on the adjoining landscape did not in reality always correspond. Views from a south-facing window, for instance, could be thwarted by the glare of the sun in 'the picture'. Likewise, eastward and westward windows in the breakfast room and study respectively might permit the warmth of the sun to enter, but the 'charm of a day-light lighted from behind the spectator can never be had'.[19] Likewise, the appreciation of views changes throughout the day, given the sun's passage and the prevailing weather of a locality. If the task, 'given a certain landscape', was to have turned it to best advantage, then one must 'comprehend the varieties of *chiaro scuro*' presented by the natural environment. Decades earlier, if George Mackenzie had been presented with so many choices in positioning his glasshouses, he might well have given up the task!

In *The Gentleman's House* the map of relations between interior and exterior spaces entailed in the book's plans and read with the author's aspect-compass assumes a guise of rationality. Its apparent objectivity belies the necessity of compromise based on the subjective values of readers and designers. The antagonism between aspect and prospect creates spaces within which one either moves or sees – at least in the mind – though in which, ultimately, residents have choices to make. The inhabitant of these spaces is cast as a sensible being. They not only occupy a particular room or spot of ground physically, but also experience precise locations visually, acoustically and even with their sense of smell. Recourse is made to an imagined experience of space through which such compromises are choreographed. Designing the home afforded 'opportunity for the exercise of much ingenuity in the disposal of rooms so as to possess the advantages of aspect and prospect together, unconnected and frequently conflicting as the demands must be'.[20]

DESIGNING FOR COMFORT

Involving the resolution of oftentimes contradictory principles, this view of design was not necessarily new to the nineteenth century. Portrayed as a kind of problem-solving, its exposition is not limited to Kerr's text. Characterizing residents in such a way as to clarify just what those principles were and how they might be addressed was novel given the growth of home-ownership, particularly during the Victorian era. It is remarkable, given the accompanying market for domestic manuals detailing the design of houses and gardens. In 1850 Edward Kemp believed that design was primarily a method of combining general principles:

Originality, perhaps, may not be deemed attainable while due regard is paid to the requirements of law. Rules are not, however, made to fetter, but merely to guide. A writer of fiction is not prohibited from representing character in a wonderfully developed and exaggerated manner. He is only forbidden from caricaturing it. Developments and extravagancies that are according to nature are in fact among the greatest merits of a work of fiction. They are at once more exciting and more elevating.[21]

The call for ingenuity by writers such as Kerr and Kemp should neither be dismissed as gratuitous nor seen as invoking simple cleverness. Rather it calls forth an imaginative process dependent on a particular way in which the home was construed as a habitat uniquely suited to human occupation. By imagining the inhabitant of plans, gentlemanly or otherwise, designers were obliged to detail specific spaces on the basis of how it might feel to live in them, move through or see through them. They were able to catalogue these spaces, label their particular usefulness or denounce them as wasted. Characterizing types of residents and reading plans made it possible to describe space using a number of subjective terms. Consider, for instance, Kerr's discussion of the necessity of comfort, and

Take, for instance, the case of a Gentleman's Study of small size; and suppose, when the occupant comes to place his desk in it, he discovers that he must choose between three evils (not an infrequent case), namely, whether to turn his back to the fire, or to the door, or to the window. He will be told, perhaps, that the reason for this awkwardness lies in the conflicting claims of a neighbouring apartment; or that it is the fault of the access, or the chimney-breast, or the prospect, or what not; but the simple fact is that it is the fault of the architect, – the room has never been *planned*. It is true, it would be dangerous to assert that the architect is bound to provide for each individual apartment an arrangement as perfect and complete as if itself alone were the subject of design; questions of compromise must continually arise, and often they will prove hard of solution; but the skill of the designer has its chief task here, in reducing every compromise, by sheer patience of contrivance, to a minimum; and the plan can never be considered perfect whilst anything of the sort is so left as to provoke the perception of a radical defect or even a serious discomfort.[22]

In this scene comfort is a measure of the position of the occupant relative to their immediate surroundings. Involving a passive response to the sensation of heat and light and the perception of spatial qualities, comfort was and remains an inherently normalizing principle.[23] It forms and directs expectations for a certain kind of experience within buildings. In Kerr's work it is related to, though distinguished from, convenience. Both result from a suitable arrangement of parts of a building to 'enable all the uses and purposes of the establishment to be carried on in perfect harmony'. In the preceding passage, however, comfort not only allows an assessment of a room's size and orientation; it articulates an individualized response to spaces and their arrangement. Illustrating another, equally subjective principle, *compactness*, Kemp wrote:

In order still further to attain the full advantage of convenience to economize space and labour, and to make everything appear orderly and well-contrived, compactness of arrangement will be particularly influential. Nothing tends more to exhibit a want of design, or to produce general slovenliness, than a scattered and ill-considered disposal of the different parts of a place. Each department that is connected with another – and all should be but parts of a combined whole – ought not merely to adjoin but to fit into its neighbouring department, so that no space may be lost, no untidy corners created, and no unnecessary expenditure occasioned on the erection of walls and other divisions.[24]

Designing for compactness could be carried to extremes, however. In advice for the intrepid house-hunter, *Cassell's Guide* (1869) warned that as for internal arrangements numerous complaints will inevitably arise upon occupancy if 'rooms are small and the ceilings low, and if the halls and passages are narrow and confined'. Imagination and foresight were required, as such uncertainties demanded a 'careful inspection'.[25]

Reading character

Conceiving of characters to inhabit their plans – or projecting themselves onto the drawing board – designers and residents alike could celebrate or condemn their houses and gardens before the first walls arose or garden beds were prepared. This seemingly minor exercise of the imagination was part of a broader historical development whereby individuals became closely identified with living spaces. This is evident in *The Gentleman's House*, which Robin Evans described as a landmark in the gradual emergence of a new subject in architectural thought. The characterization of the inhabitant during the Victorian era is distinguished from earlier rhetorical or literary conventions involving fictional personalities and settings. In the plans and paintings of Renaissance architects and artists such as Alberti, Palladio and Raphael, for instance, one discerns a polyvalent figure, free to move from room to room via multiple doors and unhindered by the restrictions of exact and conforming spaces, service rooms and auxiliary passages. It is a figure of chance encounters and animated carnality. Equally evident in literature of the period, one finds in the portrayal of manners and forms of personal engagement a relative disregard by authors for settings and furnishings and how characters might interact with them.

To the contrary, foreshadowing modern patterns of domestic life, nineteenth-century literature reveals the increasingly common differentiation of domestic spaces and their use in constructing narratives about the individuals who lived in them. In treatises like Kerr's one finds the common criticism of 'thoroughfare rooms' with their multiple entrances and the absence of hallways along with the repeated concerns for order and privacy.[26] Esther Copley's *Catechism of Domestic Economy* (1851), for instance, describes the importance of distinct 'social' apartments in the home, in which family members meet, converse and take their meals. Regardless of the social status of their occupants and the apparent salubriousness of such rooms, 'without order, a room in which every kind of work is performed is a scene of disgusting confusion and total discomfort'.[27]

Theorists have noted the influence of a range of literary works furthering this phenomenon of domestic individuation. Nancy Armstrong identifies a particular kind of power unique to women fabricated in fiction, manuals of household economy and personal conduct. In these, the household harks back to an earlier, pre-industrial world in which the family operated as a self-contained social unit. In being opposed to the world of economics and politics governing the lives of men, the household is depicted in nineteenth-century literature as a place that rose above differences between classes, acquiring a power 'akin to that of natural law'.[28] In counterpoising a world of competing interests and social constraint with the nurturing facility of the home, it was not so much the case that authors confined women within their houses and gardens. It was not that there was a place any freer elsewhere. Rather it was that the household is depicted as a fairly normal place where psychological differences between men and women or between one individual and another of either gender were most easily observed in their everyday activities and habits.[29] In either case, the ideal personal space became depicted as a projection of inner sensibilities and the home a place for restoring one's imaginative powers and motivational energies.

LITERARY TECHNIQUES

Some of the literary techniques that rendered places and habitats familiar in fiction were used by authors writing on architecture, gardening and household economy. These helped their readers relate to the rooms and landscape spaces detailed in them as though that relationship were totally 'natural'. First, the technique of using character types like Kerr's English gentleman to describe spatial experiences helped readers identify with a particular lifestyle and so bolster their own sense of belonging, well-being and moral integrity. If, in other words, a fictional character found comfort and a genial life in Kerr's well-designed study, then so too could a reader. This technique not only placed emphasis on the attributes of particular character-types like the 'soft' nature of the lady of the house, the vulnerability of children or the ribald character of bachelors. It also begged consideration of the settings in which the actions of these figures took place and of the extent to which these could be manipulated and decent habits imbued in them. Detailing the architectural qualities essential to a comfortable life, Robert Kerr described how spaciousness induces a sense of well-being in residents in the following scene:

There are many otherwise good houses in which the sense of contractedness is positively oppressive; you experience a constant fear of overturning something, a sense of being in somebody's way; you speak in a subdued voice, lest you should be heard outside, or upstairs, or in the kitchen; you breathe as if the place were musty; you instinctively stoop to pass through a doorway; you sit contractedly in your chair, and begin even to lie contractedly in bed; and to step out into the open garden, or even upon the footpath of a street, seems an act of leaping into free space! And there are others, perhaps of much less aggregate size and importance, where the mind and body, the spirits and even the self-esteem of a man, seem to expand and acquire vigour under the simple influence of elbow-room.[30]

A second, though related, literary technique is evident in the use of grammatical tense in the preceding illustration. It allowed the reader to share the same psychological space as the fictitious inhabitant of plans. After all, as the preceding passage implies, it is the *reader* who is oppressed by a sense of claustrophobia in 'otherwise good houses'. The *reader* worries about upsetting the furniture. The *reader* sits cramped in a chair or lies awkwardly in bed. Conversely, when 'amplitude of space was made the rule' as in larger or more dignified houses, the *reader* has the difficult task of 'keeping it all together'. Kerr's language is anything but plain, as he would have us believe. Rather it incorporates a set of specific meanings which both the reader and prospective designer were intended to understand. Accordingly, plans acquire an 'extended and straggling character', corridors seem 'interminable', while spaces are likely to be 'wasted'. The reader is invited to experience the spaces indicated and thereby acquire knowledge of what these terms mean when they are applied to buildings or landscape gardens.

As with language, so too the mode of spatial representation of plans requires discernment. The imaginative leap demanded of readers begins when drawings are first set before them. Kerr asked:

What, then, is a plan? Perhaps the alternative expression just employed – a map – may convey to most people the best rudimentary idea of a plan that can be had. A perfect model of the house is imagined to be constructed to a small scale – that is to say, it is measurable in Lilliputian feet and inches, the fictitious foot being perhaps a quarter, or even an eighth, of a real inch.[31]

The need to comprehend this change of scale – to envision the spatial reality represented by drawings and associated commentary – is essential for the author's arguments to seem convincing and plans at all desirable. The connection between image and reality is reinforced by diagrams like Kerr's aspect-compass. These are complemented by helpful hints urging readers to arrange small squares of paper on plans to suggest the placement of furniture in a room. By this they might understand the consequences of having too many – or too few – possessions. In an effort to impart a sense of realism to his designs for flower-beds, David Thomson suggested a similar technique:

In first considering how such a cluster of beds as is represented by this design is to be planted, and in arriving at a tolerably accurate conclusion as to the effect certain colours will produce, I would strongly recommend a simple method that I adopt, and which no brush-colouring, however cleverly executed, can approach for correctness. First, let the walks or ground-work be coloured, if it be gravel – as nearly as possible the same as it exists in the garden when in a high state of keeping, – and then colour the beds of a verdant green colour throughout. On this green, which is designed to represent the foliage of the plants, strew a few petals of the flowers, leaving green dots uncovered here and there. This will give an idea of what the plants look like when in bloom in the beds – much more correct and natural than can be given by water-colouring.[32]

A third technique of reader-identification, drawing together readers and fictional residents, rooms and gardens, called upon norms of physical habitation. Any 'normal' person would perceive and experience space in these ways. Kerr's *Gentleman's House* renders the experience of dwelling a matter of convenience and thus emphasizes thoroughfare passages and the proximity of rooms. His idealized resident is one who must mediate the demands of spaces that are moved through or seen. The reader is meant to share in the resident's domestic perambulations and obtain comfort or pleasure from them.

HOME TRUTHS AND HEROES

That literary characterization and the use of character-types should play a role in obtaining the most 'scientific' form of accommodation is notable in works of self-help like Mary Cruger's manual, *How She Did It or Comfort on $150 a Year* (1888). In a preface to the reader, the author 'wishes to say, as strongly and impressively as words can express it, that its story is not merely founded on fact, but is an actual portrayal, step by step, of her own experience, her own wonderful success in carrying out a long cherished theory of comfortable economy'.[33] Despite her claim that 'The every-day life described is not a poetically imagined affair, but one that she has absolutely lived and gloried in', the author relied on a fictional creation to argue her case. The plot involves the character, Faith Arden, as she sets about 'solving one of the difficult and perplexing social problems of the day'. Through the ensuing monologue, the reader follows the book's heroine as she takes on the task of fashioning a house on restricted financial resources. Mindful of unwarranted extravagance and meriting the determination her name suggests that she confronts her more fashion-conscious critics:

I will have a house the plan of which I have carefully studied out, in which housekeeping shall become a practical delight, with no wearying or repulsive details. I will settle down to a life of pure enjoyment, in which the grosser elements of every-day existence shall have little place. I shall have every comfort, unalloyed by household anxiety; and the bread of contentment will be sweeter to me than the richest feast you have ever spread before your guest in your own houses.[34]

Robert Kerr's character-type was implicit in a book which he wrote in 1873 specifically targeting the small country-house builder. Lest the allure of domestic riches feed an urge to 'gratify personal predilection and design a home beyond one's financial means', he suggested that:

if one feels unwilling to keep constantly before his eye the bugbear of an imaginary sale of this house, the same safeguard may be secured by regarding the building transaction as an acquisition of property by purchase. It is then obviously in his interest to get the most for his outlay, and to that end to keep always in full view the same principle of money value.[35]

The principle of economy and the character traits of foresight and forbearance are commonly prized in these books. As in works of domestic fiction, they serve to describe an ordered world, presumably under human control.[36] Buying a home was no different to selling one when it came to the kind of thoughtfulness that both activities required. Filling in time between these events, daily household management was compounded with forms of self-restraint and values permeating heterosexual relations in the family home. In what could have been a postscript to Cruger's work, Eliza Warren's tale in 1864 of managing a house on two hundred pounds a year would have been aptly titled *How She Lost It*. Her protagonist is a newlywed and the setting is the first year of a marriage. Warren chronicles a busy year, involving her character's mishandling of accounts, the selling of furniture, the move to cheaper quarters, the hiring of inexperienced help and finally the despair of the death of the couple's first and only child. These miseries were not 'God's dispensations' but 'self-created ones', in the author's estimation.[37]

STYLES OF LIVING

Before concluding this chapter, a final word is in order regarding the environs in which characters in the preceding books find themselves. In the nineteenth century an enhanced awareness of individual identity and self-worth associated with owning, furnishing and landscaping a home was allied to broader efforts to reform and improve the physical condition of the British people, their behaviour and morality. Books on residential architecture, gardening and domestic economy are notable for contributing to forms of instruction that furthered these efforts. They were published in an era in which the biological sciences highlighted the impact of the environment upon the body and mind. Biological science came to describe 'life' as a force guided by evolution and adaptation and ways of living whereby species were particularly suited to their surroundings. The characterization of the inhabitant of houses and gardens drew on broader speculation regarding the role of 'character' as that collection of personality traits, sensibilities and behaviours forming the moral fibre of individuals and nations.[38]

By way of illustration, in *The Gentleman's House* there appear several interpretations of the term 'comfort'. First and foremost, it indicates an absence of such evils as 'draughts, smoky chimneys, kitchen smells, damp, vermin, noise, and dust; summer sultriness and winter cold; dark corners, blind passages, and musty rooms'. In broader terms, it invokes the idea that each room in the house should be planned according to its purpose. Each should be 'free from awkwardness, inconvenience, and inappropriateness'.

Kerr outlines for the reader another sense of the term. He addresses the concept that comfort was part of a unique style of living. Hence, 'indoor comfort is essentially a more Northern idea, as contrasted with a sort of outdoor enjoyment which is equally a more Southern idea, and Oriental'. The reader learns that the French are motivated by certain habits that connect them to the ancient Romans. The English, on the other hand, are related to certain Gothic traits 'by direct inheritance through the Saxons'.[39] These claims of descent invoke familiar themes of 'blood' and 'soil' though they are rendered entirely relative due to peculiarities of climate, domestic habits, social distinctions and material wealth. The English lifestyle was particularly evident in the small country house, in which economy demanded judicious planning:

It is by no means an idle boast that a good English country house is the best residence in the world. Its excellence does not depend upon any merits of artistic display, for other nations are more tasteful in this respect than we. Neither is it any consideration of exceeding the luxuriousness that is in question, for we generally prefer to dispense with such refinements. The reason lies in nothing more than this: that our peculiar habits of domesticity are such, on the one hand, and our peculiar inducements of climate such, on the other, as to encourage the cultivation of quiet family enjoyment to the utmost; and that, consequently, the arrangement of our houses of the better class, especially in the country, has become a science of delightful intricacy, which when duly applied, even on the smallest scale, constitutes the edifice a thing of complete organisation, in which every part is assigned its special function, and is found to be contrived for that and no other; the express purpose of the whole being that exquisite result which is signified by our scarcely translatable phrase – *home comfort*.[40]

The idea that there are 'habits' peculiar to different kinds of people was promoted at home by Kerr and his contemporary writers and, abroad, by Hermann Muthesius in *The English House* (1908). Awareness of environmental qualities allowed readers to identify themselves with distinct modes of behaviour, manners and forms of etiquette and the places where these occurred. By the middle decades of the nineteenth century, the concept of organic structure and habitat imparted to each living being a certain distinctiveness and wholeness. Given the influence of biological thinking on other fields of inquiry such as architecture and landscape gardening, the organic structure of society was reinforced by characterizing the inhabitants of domestic space. Each was free to move between various homes, rooms and landscapes according to its own unique individual or national tendencies.

These freedoms were not without cost. Thoughts on comfort, manuals on architecture, gardening and domestic economy with their plans and techniques of reader-identification along with the goal of an ideal style of living confined all residents of the Victorian home within a complex web of obligations. The desire for comfort was accompanied by the need for a great many 'things'. It required not only spaciousness and a room for every activity, but furniture to fill rooms with. It necessitated windows to see through, but not be seen, servants who formed their own thriving 'community' behind shut doors, soil that cried out to be covered and lawns to be fertilized, weeded and mowed (see Figure 6.4). After all, the comforts of the English, Kerr advised, were due in part to their 'large share of the means and appliances of easy living'.[41] Such thoughts were built upon a burgeoning material culture and a system of production and consumption that afforded it. This point seems so obvious that little more needs to be added here. One may imagine, nonetheless, the inhabitant of Kerr's world being drawn from the web of legal obligations associated with property ownership and quartered by lines of credit, rents and mortgages, paternity and inheritance, employment and servitude. All of these were strung between the rafters of home.

As the occupant of this museum to materialism, the resident was defined in part by a need to manage its contents and appliances. He, though more likely *she*, had to distinguish between functional necessity and ostentatious ornament. They had to negotiate the breach between house and garden and equip the last wasted space. Acquiring knowledge of the attributes of the ideal home promised its mastery, but brought a transformation of behaviour as residents sought to remedy the home's deficiencies. The goal of domestic economy was family security, but it resulted in the ever more precise understanding of likely sources of anxiety and dissatisfaction. Between the lines of an ever-growing and minutely detailed catalogue of rooms and passageways with which residents of various classes and means arranged their home was written the need for discernment. It was a purposefully cultivated faculty with which the Briton cultivated sensibilities and satisfactions by knowing what to do in each of these spaces and by designing them accordingly.

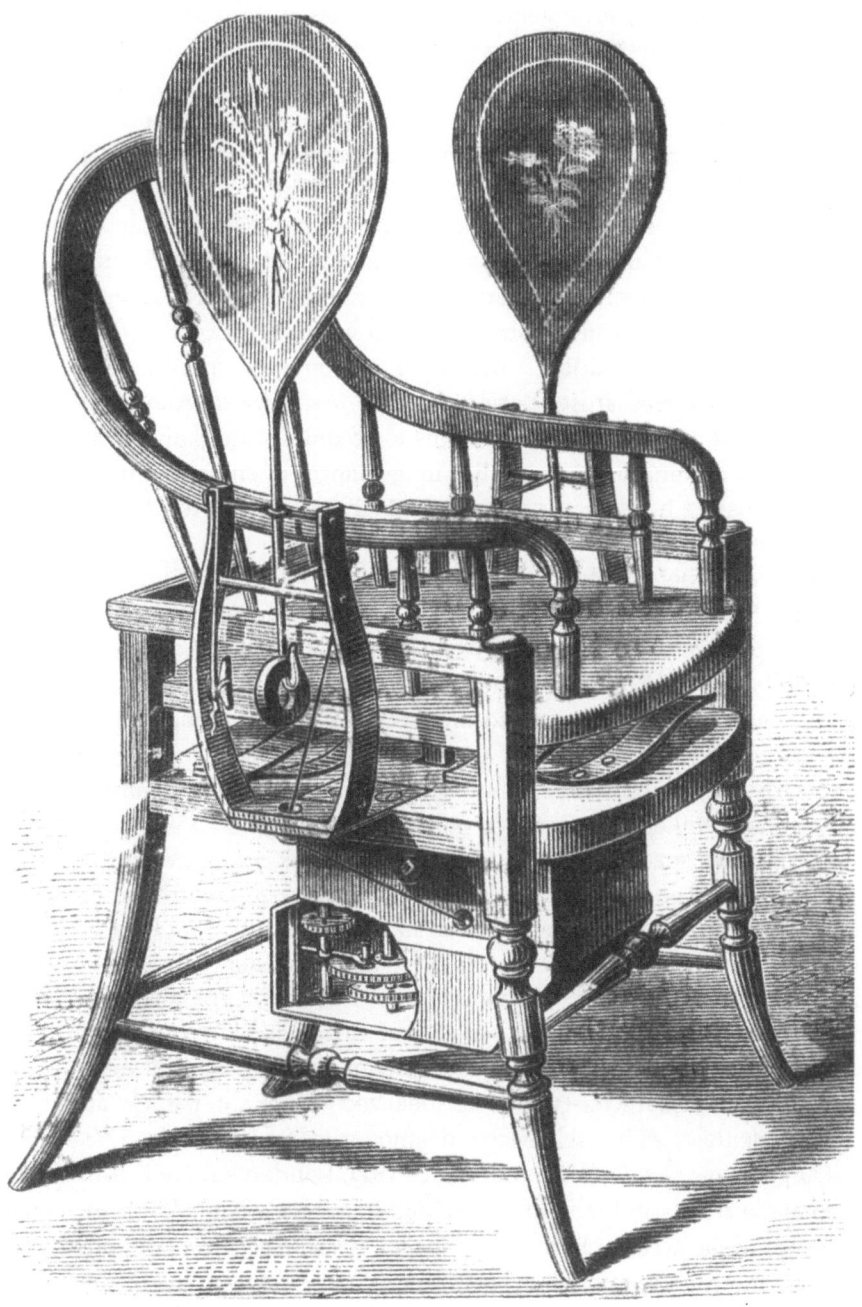

6.4 Improved automatic chair fan, 1878

The desire for comfort

Behind the gentleman's desire for a comfortable life stood a number of assumptions regarding human behaviour. Some of these were typical of any mature, healthy and fully cognizant person. They included the need for space, for instance, or a pleasing view. Implicit in Kerr's appeal to convenience and the orderly arrangement of rooms was an assumption that mind and body worked best when they worked efficiently, that is, without undue distraction or diversion or wasted effort. Other assumptions about behaviour could only be English. The need for privacy, for instance, came to be regarded as the Englishman's most cherished possession, seemingly a biological necessity, though ultimately an undeniable 'right'.

Ironically, by describing the ways any normal person, English or otherwise, would occupy and use a space, fictional inhabitants, house and garden plans could also reveal likely disruptions to domestic life. These were the ways of living suffered by any 'abnormal' person. Thoughts on the ideal home provided a basis for adding unusual or recuperative spaces to its catalogue of rooms, as need arose. Designers could rearrange thoroughfare passages or reconsider the proximity of rooms to accommodate individuals somehow crippled in mind or body by the circumstances of environment or fate. They could make a house or garden *so* private that one could retire from society completely.

Twenty-six years before the first edition of *The Gentleman's House* appeared, John Loudon described the plan of a house and garden in his equally encyclopaedic book, *The Suburban Gardener and Villa Companion* (1838). Loudon was 55 years old at the time. The plan was purposefully conceived to 'have some pretensions to architectural design; being, at the same time, *calculated for invalids*, and, therefore, furnished with verandas extending nearly round the whole building for taking exercise in during inclement weather'.[42] Like Kerr's terraces and parterres, Loudon's verandas form an intermediary space between rooms and the outdoors, between the comforts of the former and exposure to the elements and neighbourhood of the latter. The representation in plan of this architectural feature does little to suggest the quality of the space it encloses (see Figures 6.5 and 6.6); rather, plan, elevation and axonometric drawings show relationships between spaces designed for very specific purposes: ease of movement and protection from inclement weather. In Loudon's design convenience is enhanced even further for the specific character-type of *his* idealized inhabitant, namely an invalid. In real life this was himself. Melanie Simo writes that at the time his house was largely completed, in the autumn of 1823, Loudon suffered from rheumatism and a right arm that had broken and improperly healed. It is an ironic postscript to the account here of curvilinear glasshouses that Loudon, a figure most concerned to provide a nurturing enclosure for plants in the forcing-house, ended up in one as well. This was given the form of a small, domed conservatory placed at the centre of the street façade of his house.[43]

In both Kerr's house for a gentleman and Loudon's home for a rheumatic plans and associated commentary allow for assessments of normal ways of living and those that are slightly less so. Both diagrams show ways of using, moving through, visualizing and obtaining comfort from the room they represent. Relations of proportion and other ambient qualities are subordinated to those of proximity and the interconnectedness of space. The

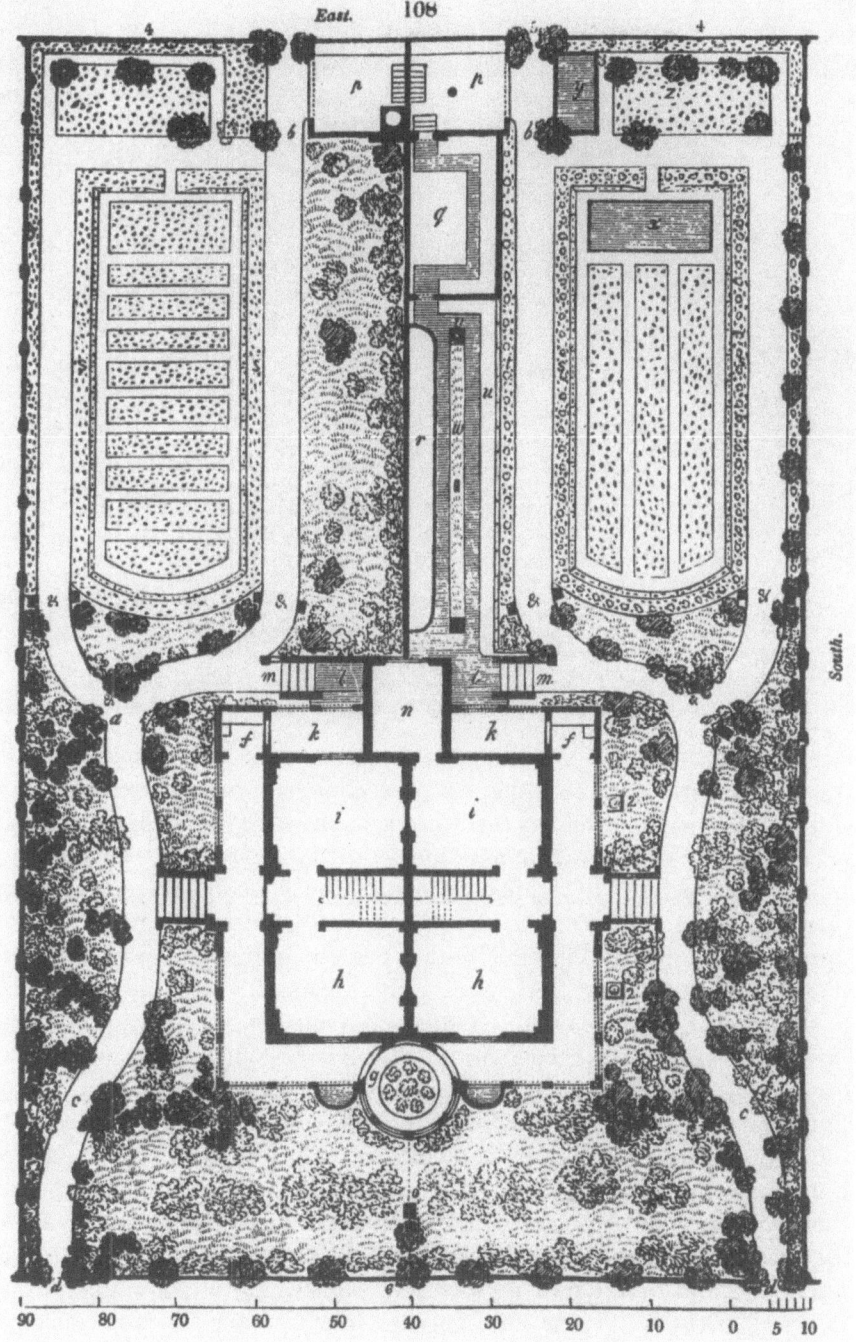

6.5 'A double-detached suburban villa plan in Porchester Terrace, Bayswater', London; in John Loudon, *The Suburban Gardener and Villa Companion* (1838)

6.6 John
Loudon,
streetside view
of the 'double-
detached
suburban villa'
at nos 3 and 5
Porchester
Terrace,   ,
Bayswater,
London, c.1823

labelling of interior rooms and garden terraces is keyed to their respective purposes. Architectural style is of secondary importance to a more fundamental quality, namely the organic wholeness of the built environment. Plans and the text accompanying them provide the basis for knowledge of the specific nature of human habitation by functionally differentiating between rooms and by describing relations between them. In terms of broader assumptions regarding human behaviour, this facilitated an overlap between the fully integrated building and the morally integrated life.

Ironically, constraints of possession, management and interpretation associated with home-ownership in Victorian Britain failed to supplant an overarching belief in domestic freedoms. Kerr described the home as the Englishman's 'most cherished possession'. It was occupied by a species that wished to 'avoid obtrusiveness' by its very nature.[44] Given similar claims in numerous books of the period, the Victorian assumed the guise of the ideal householder and homemaker. The freedoms entailed in health and well-being were freedoms from draughts and fetid air, from the dampness of walls and linen, and from dark, cramped spaces as well as unwarranted physical and visual contact. These little liberties were exercised by first imagining characters to inhabit plans and designs, abstract figures subject to both domestic obligations and a right to obtain shelter and find a little 'elbow room'.

It is interesting that Loudon chose his own house – and by implication the circumstances of his own life – to illustrate principles of planning and good design, thereby making a lesson of himself. The use of characterization to configure the Victorian home coincided with a change in relations between designers and their clients. Both Kerr and Loudon were designers, critics and educators of a kind. Both stressed the need for methods to design house and garden more scientifically. It was no longer acceptable to copy the work of Palladio or his eighteenth-century admirers. Classical values remained important, though direct communication with broader classes of readers and clients was necessary to address their own circumstances and needs – their own 'spot of ground'.[45]

Kerr's authority served to articulate and represent to the householder or prospective resident a 'psychopathology' of the home.[46] Based on the model of contagious disease, this condition depended upon the resident's self-conscious experience and interpretation of their surroundings. It placed a moral imperative on each person to remedy the diverse and environmentally determined causes of stress in their lives.[47] These varied from individual to individual, between man and woman, and so houses and gardens varied in their detailed planning and design.

The categories of convenience and comfort, so central to Kerr's *The Gentleman's House*, were not states imposed upon the individual. They were not simply the effect of particular environs, a specific configuration of rooms and corridors or relation of aspect and prospect. Rather, they were dependent upon a particular kind of living being of greater or lesser degrees of normality. It was a being endowed with freedom of choice and a readiness to assume a lifestyle responsive to their own needs for health and well-being, familial cohesion and emotional stability. The desire for domestic integrity, like the 'duty to be well' in the context of biology or medicine, depended upon visible signs of 'initiative, adaptability, balance and strength of will'.[48] As a result, the domestic environment was typecast as a series of scenes for the enactment of one's moral life.[49] The Victorian house and garden provided an arena for various bodily habits, family relations and modes of individual expression. Though connected to the world of labour, they encouraged disciplines and creative practices that were not totally determined by economic concerns, but were rather, somehow, more intrinsically human. For the social theorist, the Victorian home was a site where powers governed at a local or intimate level. It was a place where medical, psychiatric and educational theories informed a range of bodily and building practices. One must be mindful, however, that the Victorian home was also a place for imagining what it meant to be a unique person, to be blessed with autonomy and personal freedom.

Memory and the garden cemetery

In the gentleman's house – or, more generally, in the study, the veranda, the garden or parterre – spaciousness may have secured a sense of freedom, but time nevertheless proved to be a source of constant anxiety. Spatial acuity is one means of describing the uniqueness of human beings. It is a key facet of human experience, shaped by the developments which accompanied industrialization and urbanization as well as growing environmental awareness. An appreciation of time is another such facet.[1] The exploration of novel ways of accessing the past, of managing the present and affecting the future supported the redefinition of human values in nineteenth-century Britain. Perhaps never before had so much attention been given to how urban and domestic spaces and landscapes might accommodate individual lives to the duration and finality of life, to desires for regularity, order and security and to the prospect of a nurturing home. Over the course of the century, the built environment came to memorialize, house and enhance sentiments, qualities and values construed as indivisible parts of human nature. Some of these, such as sorrow, self-assurance, innocence, were subjects of longstanding philosophical speculation. They were transformed, however, in light of new scientific discoveries and moral psychology, to serve new ways of understanding and governing society as a whole.

The sociologist Norbert Elias described time-consciousness as an 'all-embracing compulsion' entailing both a form of essential orientation and a means of self-control.[2] In his view the purpose of this compulsion was to channel instinctive drives to serve communal purposes. Time and other symbolic systems were means for the reconciliation of individuals to both their natural and social surroundings. Such systems are individuating, serving to articulate individual differences and to provide the basis for shared ideals. This view of time, serving a functional or practical purpose, is seen to be espoused by some social theorists. These systems also serve, like broader appeals to language, to emphasize the uniqueness of humankind based on a supposed, innate capacity for self-awareness and control. Like language, time-awareness has served as a model of consciousness and self-reflectivity, being a means to accommodate change.

It remains a common enough claim that one can learn from the past or that there is no time like the present. Time and the functional and symbolic aspects of one particular nineteenth-century landscape came together in the belief that human capabilities could be extended and improved. In garden cemeteries, proposed and built in Britain in the early decades of the nineteenth century, awareness of time's passing was purposefully cultivated as the basis of individual character. This occurred alongside broader theoretical developments associated with the emergence of the life sciences, including studies of memory and cognition, attempts to discern the psychological basis of personality and investigations into adolescent behaviour and development. These studies attached new meanings and significance to the continuance, force and development of life itself.

A silent and decent abode

An awareness of time was implicated in the expansion of biological science with the dawning recognition that life is dynamic and gives rise to organic forms that develop through some kind of adaptation to diverse habitats and ways of life. This idea was given its fullest formulation in the work of Charles Darwin, though the likely existence of a universal process of organic evolution was admitted earlier by other writers, most notably by Jean-Baptiste Lamarck (1744–1829). Crucially, by placing life processes within a historical context, their works supported new temporal frameworks that came to articulate biological facts upon a field of social relations.

One clear instance of this is evident in the shifting responses to the fact of human transience. Mortality, of course, has long shaped human sentiment. However, an awareness of the regular gain of births over deaths in many parts of eighteenth-century Europe helped shape a new social reality: population.[3] Robert Malthus's much-cited *Essay on the Principle of Population* (1798) is notable in establishing a direct relation between rapidly expanding population and diminishing resources. His study formed a calculus of consumption and established population growth as a key vehicle of social change. Similarly, a boom in statistical analysis in the following century and the study of longevity rates made it possible to relate measures of the duration of life to the degree of healthfulness of rural districts and urban quarters.[4]

Observations of the high incidence of disease in areas adjacent to Britain's burgeoning cemeteries in the 1830s and 1840s prompted the reform of burial practices in the nation's cities. Reform was accompanied by an avalanche of numbers, detailing birth and death rates, rates and densities of interment, incidents of cholera and other suspect diseases attributed to the inhalation of decomposition gases. Numbers drew reformers and Parliamentarians to the grave. In addition to their concern for the unhealthy effects of crowded and indiscriminate interments, writers bemoaned the loss of opportunity afforded by a well-designed necropolis, the deprivation of so much 'gratification' and 'profitable instruction'. By initiating the garden cemetery movement, reformers turned death to democratic ends. They argued for a type of nature reserve where one could observe 'those who on earth held most different religious opinions lying together in one space, one close, and as it were in one flock, waiting for one Shepherd'. In the garden cemetery the past became a social fact of a uniquely human experience of life. Ever mindful of the shortness and uncertainty of life, but introduced to spaces in which time's passing was organized and made purposeful, visitors to the cemetery were thought to better orient themselves towards tasks that lay ahead.

An event that was a catalyst for new modes of burial in Britain was the establishment in 1804 of the Père-Lachaise cemetery on the slopes of Mont-Louis at the northern edge of the city of Paris. Following the publication of accounts extolling the virtues of the necropolis, critics of garden design and architecture joined ranks. Along with the clergy, social reformers and philanthropists, they called for the construction of new model cemeteries for Britain's major cities. To the goal of providing a hygienic alternative to the putrefying masses growing beneath the metropolis was added an expressed desire to provide the working classes with lessons drawn from past lives. These were to be entailed in the carefully contrived appreciation of monuments to famous, though more commonly industrious, countrymen.

In 1831, John Strang, a City Chamberlain of Glasgow and a figure closely associated with the cemetery reform movement there, observed that the lessening of tensions between Great Britain and its continental neighbours after years of war brought a greater willingness to follow the European precedent where reason triumphed over sectarianism and national antipathy in the 'administration' of the dead. To the improvement of his homeland and its cities he urged fellow citizens to imitate the French and build a radically new kind of garden cemetery, a 'Scottish Père-Lachaise'. This proposal led to the creation of the Glasgow Necropolis on a hill adjacent to the city's cathedral (see Figure 7.1). Strang intended that:

In that vast grove of the dead, each has his own grave, and each his own mausoleum. In place of the clumsy mound or large white stone, that so generally covers the ashes of our countrymen, is to be found a little flower-garden, surrounded by cedar, spruce, cypress and yew trees, round which the rose and the honeysuckle are seen entwining, while, instead of a solitary and deserted church-yard, the eye meets at every turn with some pensive or kneeling figure weeping over the remains of a relative, or worshipping his God at the tomb of excellence and virtue.[5]

Few reformers and landscape gardeners failed to acknowledge the beauty of the Parisian cemetery. George Collison's *Cemetery Interment* (1840), for example, entails a history of burial practices. It relates Père-Lachaise to ancient customs, to provincial and overseas burial grounds and to the most recent of the new institutions in London, Abney Park Cemetery, which was established on the outskirts of the metropolis at Stoke Newington one year earlier.[6] In the United States lingering revolutionary fervour led the leaders of the rural cemetery movement there to propose a new burial ground at Mount Auburn in Philadelphia. It was to be named the 'American Père-Lachaise'.

7.1 Glasgow Necropolis, contemporary view. Note that the churchyard of Glasgow's cathedral in the foreground is literally paved over with tombstones, evidence of the crowded conditions in traditional urban burial grounds throughout Britain that were much criticized by reformers. The necropolis sits on the hill in the background – a healthy, 'spirit-stirring' alternative

Other, less jingoistic reformers successfully argued that the gates at Mount Auburn should carry the more pedestrian title of 'Monument Cemetery'. They noted that its internees were celebrated for having 'rendered blessings to mankind', not only to Americans.[7]

Histories of the rural cemetery movement in America described the role played by such places in countering a then prevalent and typically British view of America as a cultural wasteland. Citing the popularity of Mount Auburn among local and foreign tourists from its establishment in 1831, Stanley French has described the dual purpose of such institutions as both pleasure-gardens and cultivators of public spirit.[8] One particular Swedish visitor at the time is recorded as declaring 'a glance at this beautiful cemetery almost excites a wish to die'. Whether this serves as comment on the reality of death in early nineteenth-century America or life in Scandinavia at the time is open to interpretation.

In both Britain and the United States the cemetery's role as a space for celebrating the past and for encouraging personal remembrances and sentiments was to acquire a unique form, given the appreciation of enterprise and hard work shared by the citizens of these nations.[9] Among the Anglo-Saxons, the garden or rural cemetery came to play a prominent role in the pursuit of self-improvement. There, the appeal to the fecundity of nature as not only affording solace for the bereaved, but also cultivation for the living drew renewed attention to the design and placement of monuments in garden-like settings and the careful arrangement of paths and woods.

Accordingly, in what has become a familiar stereotype of continental ostentation, reformers were likely to criticize Père-Lachaise for its opulence and extravagant ornamentation. The celebration of art over nature led one American commentator to observe:

in Père la Chaise every expression of mourning is to be found; few or none of hope ... There is no light from the future shining over the place. In Mount Auburn, on the contrary, there is nothing else. A visitor from a strange planet, ignorant of mortality, would take this place to be the sanctum of creation. Every step teems with the promise of life.[10]

Acknowledging the notoriety of Père-Lachaise in his own monumental work, *An Encyclopaedia of Gardening* (1822), John Loudon agreed with a contemporary's observation that the cemetery was 'tricked out and overacted, as if there were nothing sacred from impertinence and affectation'.[11] He noted the great expense of burials resulting from such extravagance. More notably, natural defects of horticulture were aggravated by the Gauls' 'excess of art' by 'too many walks; by too many seats and buildings; and by too meagre a distribution of trees and evergreen shrubs'.[12] Loudon's criticism of the extravagant use of monuments and buildings in Père-Lachaise was extended to include the insufficient and ill-advised plantings that he believed characterized French gardens in general. In one particular park, for example, sparingly though 'scientifically sprinkled with wood', the multiplicity of walks brought confusion to the eye of the spectator rather than 'grandeur, richness, and repose'.[13] The result of such immoderation was a disorder that numbed the mind to intended associations and sentiments.

The call for restraint in design and meaningfulness in scenic effect in the new British cemetery suggests parallels to pedagogic concerns appearing at various times in the first half of the nineteenth century. It calls to mind the debate leading to the establishment of a national education system in Britain to follow Scottish and continental precedents. Consideration of these developments has been anticipated in previous chapters. Concerns included thoughts on the usefulness of one taxonomical system versus another among botanists and the usefulness of learning in archaic languages more generally. They raised questions as to the best method to arrange encyclopaedias or books in libraries, artworks in the museum or plants and animals in the botanical garden or zoo. In each case the careful arrangement of the objects of knowledge was seen to be central to the formation of knowledge itself. If only one could devise a 'better' way to arrange specimens, books and monuments to the past and then provide space to view them, it was commonly thought that learning would proceed more effectively (see Figure 7.2). In the public sphere, individual character and behaviour – whether of the mournful subject, the diligent pupil or the respectful observer of culture or nature – might likewise be improved.

It was hoped by reformers involved in the garden cemetery movement that, if tombs and their epitaphs were dispersed among botanical specimens

7.2 Castings gallery, Victoria and Albert Museum, London, contemporary view. The castings were made of key works of European architecture and sculpture and organized to illustrate 'northern' and 'southern' European influences. The collections were intended to form a library to be shared among nations and to illustrate to the artisan and lay-person the best works that civilized nations had to offer posterity

systematically labelled and arranged, the cemetery would become a means for promoting civil obedience. The graveyard would not only serve practical purposes but promote good citizenship among the masses. Cemeteries, like schools, libraries and museums, provided so many 'silent and decent' abodes for reflecting on the consequences of one's own life. They were places for thinking about the progress of the British people. In such spaces the exercise of memory and solitary pursuits became integral and integrating aspects of human experience. They were intended to be beneficial to the population as a whole, not solely to philosophers and aesthetes.

Burying people and raising citizens

In 1829, twelve years before the publication of his treatise on laying out, planting and managing cemeteries, Loudon devised a 'Plan for a National Education Establishment'. It was a proposal aimed at children from all classes of society, from infancy to puberty, and offered an alternative to instruction by the national churches of England, Scotland and Ireland. At the outset of his thesis, Loudon asserted that 'all human happiness and prosperity, whether public or private, domestic or national, was founded on individual cultivation'.[14] The new model cemetery and classroom shared similar purposes. Both were to mitigate the differences between rich and poor, as Loudon wrote: 'Supposing education to be a fluid, we would immerse every male and female child in it, not only for the same length of time, but in order to let the rich become personally acquainted with the poor, and the poor with the rich, *in the same vessel* [original emphasis].'[15] Cemetery reformers and educationalists alike further sought to lessen differences between the rural labourer and urban literati:

The country churchyard was formerly the country labourer's only library, and to it was limited his knowledge of history, chronology, and biography; every grave was to him a page, and every head-stone or tomb a picture or an engraving. With the progress of education and refinement, this part of the uses of churchyards is not superseded, but only extended and improved. It is still to the poor man a local history and biography, though the means of more extended knowledge are now amply furnished by the diffusion of cheap publications, which will at no distant time, it is to be hoped, be rendered still more effective by the establishment of a system of national education.[16]

Behind calls for a national system of education and the reform of burial practices was a general view that the established Church was ill suited to provide moral guidance to a large and increasingly diverse population. It was unable to solve the many practical problems associated with the population growth in metropolitan areas. In Britain, as had occurred earlier in France, governments, through their own actions or those of regulated private companies, were called upon to establish a medium for diverse expressions of religiosity as an adjunct to improving hygienic conditions relating to burial.

Not everyone agreed with proposals that the burial of the dead should be left to private enterprise. Vehement protests arose from priest and grave-digger, both hitherto responsible for interment and deriving from the trappings of death a considerable portion of their income, if not all of it.[17] Nevertheless, in the public cemeteries built in the 1830s and 1840s, distinctions between the burial grounds for Christian and Jew, Catholic and Protestant, Anglican and Nonconformist became less commonly the result of mandate by the established Church or act of Parliament. Rather, they followed individual preferences and tastes with options for differing modes of burial promoted by newly established cemetery companies.[18] These purportedly 'democratic' – and highly profitable – institutions created places where, according to John Richards (1843):

To behold those who have differed while living, neighbours in the grave, cannot but whisper a lesson of a character eminently fitted to have a beneficial effect on the conduct. 'There is neither speech nor language,' but yet there is a voice in such dread retreats of an assuaging nature, tending to soften men's manners, and to instigate charity and humanity. The consecrated will be a defence and guard to the unconsecrated part; and the unconsecrated, embracing in a manner the consecrated part, must naturally promote kindly feelings. As in the natural world by contact, the most discordant materials produce beneficial effects, so there is nothing in the moral world that would make us augur anything but good results, from this concentration of sepulture.[19]

In assuming responsibility, directly or through commercial licensees, for educating the masses, as well established in several German states by 1829, and for burying them in Britain a decade later, it was hoped that the governments would reap the benefits of reduced religious animosity and heightened civic virtue. Consequently, greater social stability would follow. In achieving these goals an ensemble of pedagogic practices, exemplary figures and novel spaces relied on a particular understanding of character as that quality of the individual that was both existential and ultimately malleable. They likewise relied on an understanding of the past as offering a vast storehouse of valuable lessons to be had through attentive study.

Loudon's treatises on education and cemetery reform prefigured arguments put forward by educationalists such as Joseph Lancaster (1838) and David Stow (1850).[20] These works were similar in that they both offered, as the title of Stow's work suggests, a comprehensive and detailed 'Training System' or method of instruction. Such schemes depended on the moral example provided by well-chosen teachers and headmasters as well as on the architecture of the classroom and landscape of the school-yard. These offered means of surveillance and the orchestration of self-expression through study and 'freedom' of reading and play. In Loudon's system, the school-house was to contain no fewer than five rooms, one of which was to be a library and museum. Others were dedicated to the instruction of boys, girls, infants and the adult population (in the evenings) respectively. The school-yard was a botanical garden. The design was left up to each parish, provided it was at least an acre in extent and exhibited and explained twelve of the Linnaean classes and the leading subdivisions of the Jussieuan classes – the two leading botanical classification systems of the day.[21] Loudon's school-yard calls to mind his project for a public arboretum at Derby (1839). There, the combination of recreational pedestrianism and the study of nature arrayed according to genus and species was likewise thought to have a positive influence on public morality. It also had something in common with his more grandiose schemes for the Birmingham Botanical Horticultural Garden. These projects were all, in a sense, 'living museums'.[22]

In Loudon's plan, the study of botany by school children demonstrated links between living forms and their purposes. This instilled an appreciation of the overall order of the natural world. The desirability of one classification system over another and the need for fluency in Latin to decipher taxonomical terms were less important as issues or skills in themselves. Rather, they were handy as tools affording an understanding of nature's fullness and interconnectedness. Inside the classroom, the study of English literature would provide a means for appreciating social order, for demonstrating links between fictional characters and forms of desirable behaviour and moral values that could be acquired by adolescent readers.[23] Loudon was to join a number of educationalists who denounced instruction in the 'dead' languages associated with Classical and, subsequently, Romantic aesthetic education as well as the elite form of erudition such study entailed.

Developments in pedagogy by the middle decades of the nineteenth century brought into play a new concept of human identity, where 'character' informed an understanding of personality and behaviour. Recall that Robert

Kerr used a form of characterization or literary deceit to detail the manner in which one moved through and derived comfort from the home. Its use in a text such as *The Gentleman's House*, serving largely to accommodate individuals to their surroundings, performed a different role to that used in earlier works, where literary character appears as one of a number of elements, including plot and setting, from which dramaturgical conventions were derived and forms of behaviour were rendered distant and abstract.

In the genres of eighteenth-century elegiac and consolatory poetry, for instance, a universalizing of character is found in the voice of the obedient Christian or reflective melancholic. In such works as Robert Blair's didactic poem *The Grave* (1743) and Thomas Gray's famous *Elegy Written in a Country Churchyard* (1750) death is depicted as life's only certainty, the Great Leveller respecting neither age nor rank (see Figure 7.3). While the reader is meant to identify with the sentiment, its narrator escapes scrutiny, being represented neither as a 'real' person whose life history is a lesson on the role of past circumstance or environment in the determination of behaviour, nor as an individual in possession of specific social characteristics and moral attributes. The reader learns of such characteristics and attributes – if at all – largely through subsequent biographical detail.

James Stevens Curl has suggested that Edward Young's famous elegiac poem of 1742, *The Complaint: or, Night Thoughts on Life, Death, & Immortality*, contributed to the origins of the garden cemetery, when the historian noted a departure from the Christian concept of eternal life to a growing emphasis on the value of the past, memory and commemoration.[24] Curl draws attention to the international significance of the poem, due partly to the quality of its verse, and cites the following passage as typical:

> *This* is the Desert, *this* the Solitude;
> How populous? how vital, is the Grave?
> *This* is Creation's melancholy Vault,
> The Vale funereal, the sad *Cypress* gloom;
> The land of Apparitions, empty Shades:
> All, all on earth is *Shadow*, all beyond
> Is Substance; the reverse is Folly's *creed*;
> How solid all, where change shall be no more?[25]

Contrast Young's poetic musings, however, and the learned audience for which they were intended with the moral proselytizing of a much-cited work like John Strang's *Necropolis Glasguensis*, written almost one hundred years later:

A Garden Cemetery [wrote Strang] is the sworn foe to preternatural fear and superstition ... A Garden and monumental decorations are not only beneficial to public morals, to the improvement of manners, but are likewise *calculated* to extend virtuous and generous feelings ... They afford the most convincing tokens of a nation's progress in civilization and the arts ... The tomb has, in fact, been the great chronicler of taste throughout the world.[26]

A chasm of meaning, highlighted by Strang's calculus of sentiment, opens between the Romantic's predilection for odes and funerary architecture and

7.3 Illustration by Harry Fenn, in Thomas Gray, *Elegy Written in a Country Churchyard* (1884, ₁1750)

the reformer's concern for public welfare, which an assessment of literary sources behind the garden cemetery movement fails to bridge. The garden cemetery transformed the raw material of sentiment and its expression in earlier forms of literature into a conciliatory apparatus and means of self-improvement. Its basic alphabet, observed by Strang in Père-Lachaise, consisted of: 'a square or parallelogram of ground, of about three or four yards broad, enclosed by a neat little railing of iron or wickerwork. Within this spot, there is always either a sepulchral urn, a small pillar, or a cross, to tell the name and the quality of him who lies below.'[27]

The project to understand the past in the cemetery and to learn from the lives of characters buried there led to an interpretative schema composed of basic units such as these. Each illustrated moral truths and desirable forms of behaviour. Just as Kerr's portrayal of the retiring gentleman invoked specific ways of moving through, visualizing and obtaining comfort from space, character-types and life histories of those ultimate retirees – the dead – proved exemplary in guiding the living to a 'better' end. The characterization of the dead served pedagogic practices aimed at transforming individuals into better-behaved, responsible, or more productive members of society. For Richards, the cemetery was a 'contrivance for gaining more life'.[28] In praising the poets, artists and philosophers of European but, more tellingly, British culture, Strang acknowledged common sentiment, but eulogized the work of exemplary lives:

It is a singular fact, that the most sublime thoughts of man have almost always sprung from under the weight of affliction. It was in the presence of death, it may be said ... that Gray conceived and penned his immortal Elegy. It was in the prospect of a scaffold, that Sir Walter Raleigh completed his 'History of the World,' and it was when he stood on the very verge of time, that he produced some of his most affecting verses. It was while bewailing the death of a fond mother, and for the pious purpose of raising money necessary to defray the expense of her funeral, that Johnson wrote his Rasselas; and it was while wandering through the magnificent cemetery of Bologna, whither, during his sojourn, Byron almost daily bent his footsteps, that the noble Poet cherished the powerful fancy, which in his dreams of 'Darkness,' conjured from their tombs the spirits of the past, and introduced the living to the generations with whom they are destined ere long to be actually associated![29]

The cultivation of enterprise

In its modern form, the concept of character suggests qualities that are somehow unique to the human species and to each of its individual members. The concept owes much to the rise of life science in the nineteenth century and, ultimately, theories of the mind that sought to address human identity in new ways. An understanding of character was affected by studies of moral psychology and theories of race so that entire peoples were said to possess a unique identity and individual members of a race or ethnic group specific qualities of mind and spirit. Samuel Smiles believed this and readily asserted in 1859:

One of the most strongly marked features of the English people is their indomitable spirit of industry, standing out prominent and distinct in all their past history, and as strikingly characteristic of them now as at any former period. It is this spirit, displayed by the commons of England, which had laid the foundations and built up the industrial greatness of the empire, at home and in the colonies.[30]

Loudon described one of the 'great evils' afflicting British society as the over-production, not only of manufactured goods, but also of human beings.[31] By educating the masses he believed the Government would encourage respect by even the privileged classes for the prosperity of *every* citizen as well as appreciation of the means of achieving wealth by the poor. Encouraging a value for education would cause the labouring classes to defer from marrying young or from having too many children, lest insubstantial financial means deprive them of suitable training and the means of security. An additional benefit of forbearance would be that the Government and charitable organizations would have fewer hungry mouths and idle citizens to satisfy. This was a lesson apparently lost on the Chinese in Lord Macartney's day. Better to have an 'enlightened superfluous population' wise enough to emigrate and find fortune in other countries, such as Australia. It is perhaps worth wondering, along with Loudon, just 'How is it that there are more Scotchmen to be found in other countries than either Englishmen or Irishmen? It is simply because they are taught a little more at school.'[32]

The extension of the British Empire would absorb the energies of this excess in humanity just as the earth of the garden cemetery was intended to absorb their spent bodies. A curious understanding of the purpose and social context for work – of a kind to be expected from an anthropologist today – and the sentiments, motivations and social organizations that support it, appears in the treatises cited so far. It is evinced by the histories of memorials, their makers and the burial practices of other peoples that commonly introduce many of the tracts of cemetery reform (see Figure 7.4). John Richards chose a familiar starting point in his history of 1843: 'Those stupendous works of art, and most wonderful specimens of mechanical genius, the Pyramids, show that the ancient Egyptians knew how to honour the dead. Having survived numberless generations, they still lift their lofty heads unscathed, whispering a wholesome lesson to the hearts and consciences of mankind.'[33]

Strang pointed out the centrality of the tomb to Egyptian society and eulogized this 'cradle of philosophy, of science, of art, and of legislation', where 'veneration for the dead was carried to the highest pitch'.[34] Almost thirty years previously, Quatremère de Quincy expressed equal appreciation for the distinctiveness of Egyptian architecture, describing it as inherently tomb-like as opposed to the huts of the Greeks and the impermanent dwellings of Chinese society. Strang described the terminal, rather than the starting, point of Quatremère's 'epigenetic' theory of the evolution of architectural forms. His references to the past served largely rhetorical purposes in criticizing contemporary horticultural designs and inadequate burial practices. This was obviously the motive behind his call for a Scottish Père-Lachaise to equal the efforts which were taken abroad. Given the valuable lesson left by Egyptian, Jew, Greek, Roman and 'other nations from the earliest ages', it became the duty of society to care better for those whom John Richards called its 'defunct'. In Richards's view, given that even the 'rudest' of nations and most 'savage tribes' – even the French – exercised this universality of sentiment, then should not Britain as well? It was obvious that an industrialized nation like the United Kingdom required a vehicle of commemoration suitable to its time, for:

Will it not rather be gratifying to us to give our best support to a project embracing such objects? It would indeed be very reprehensible, if when hospitals, Athenaeums, museums, guildhalls, bridges, aqueducts, viaducts, railways, and even gaols, are appearing in every direction under the hands of taste, science, and art, we should leave the dead to be buried in places hardly fitted in many instances for the interment of a mule or a dog.[35]

The valorization of nature, so central a reference in histories of the garden cemetery movement, played a new and important role in the anthropology of work invoked by Richards and Strang. The Romantic predilection for nature was reinvigorated in the nineteenth century as an important aspect of universal human behaviour. Nature was an essential vehicle for the recuperation from grief and the restoration of motivational energies. Collison found in the solitude of woodland or calm of pastoral scenery the means by which the troubled mind

7.4 'Cemetery of Eyub, near Constantinople', in John Claudius Loudon, *On the Laying Out, Planting, and Managing of Cemeteries; and on the Improvement of Churchyards* (1843)

was satiated.[36] Trees provided an ideal model of patience and forbearance for the grief-stricken. They provided shade, ameliorated soil and climate in terms reminiscent of an accountant's balance sheet, while 'year by year improving, at more than compound interest, the value of their original cost, the labour of their planting, their occupation of soil'.[37]

In these tracts, metaphors describing the fecundity of the soil stand in stark contrast to the horrors of earth rendered noxious by incessant burial. At the outset of his plan for managing cemeteries, Loudon described his primary object to be the return of the remains of the dead to the earth 'from which they sprung', a familiar Christian sentiment. His secondary purpose, that of improving 'moral feeling', of rendering the burial ground a 'source of amelioration or instruction' was dependent upon making the cemetery attractive:

So far from this being the case at present, they are in many instances the reverse, often presenting, in London and other large towns, a black unearthly-looking surface, so frequently disturbed by interments that no grass will grow upon it; while in the country the churchyard is commonly covered with rank grass abounding in tall weeds, and neglected gravestones.[38]

Abhorrence of corruption in the churchyard guided Richards's call for reform, arguing that their 'soil' nonetheless remained

congenial, and a little cultivation is only wanted to ensure an abundant crop. So convincing, indeed, are the arguments in favour of these undertakings, that a little while, and the mists of opposition must be dispelled, and the people will be ready to call down a blessing upon those who designed and have prosecuted to completion such good works, such labours of love, such realisations of god-like energy.[39]

By the time Richards's *Essay on Cemetery Interments* appeared, humankind may have then fallen from its privileged position into a morass of chemicals and bio-molecular processes. Nonetheless, it mattered to the author what one did with one's own 'matter' while alive. The ground of the cemetery was a resource for the manifestation of 'god-like energy' in many forms. Paradoxically, it was a record of the vitality of life itself, both literally as terrain to be used and rehabilitated and symbolically as a place to return to and to identify with as a place of origin.

Evoking another dimension of this culture of enterprise was the representation of the cemetery as the reserve of nature, a place protected from human interference or degradation. Even before 'the subsoil began to be fashioned into bricks, and the builders commenced to eat up the land and transform trees into houses', places like Abney Park in Stoke Newington were associated with important people and significant events, and the cemetery preserved these memories and added to them 'year by year' (see Figure 7.5).[40] In the choice of landscape styles for the garden cemetery, a preference for familiar 'natural' features over highly contrived ones accounts for the favouring of winding paths and irregular terrain commonly identified with the English picturesque. These were opposed to the rigid formalities associated with the *ancien régime* in France.[41]

Contrivances for life

The historical circumstances that guaranteed the cemetery its exemplary status included forms of expert knowledge, the particular planning and detail of the cemetery itself and a specific way of reading the landscape. Among other forms of discernment, this was based on the appropriateness of architectural and garden features for their time and place. For Strang, while the tomb may have, 'in fact, been the great chronicler of taste throughout the world', it was only one of many forms for expressing the universality of grief. It was likewise an important marker of cultural distinction:

is it not on the urns and sarcophagi of Etruria, that the lover of the noble art of sculpture still gazes with delight? And is it not amid the catacombs, the crypts, and the calvaries of Italy, that the sculptor and the painter of the dark ages chiefly present the most splendid specimens of their chisel and their pencil? In modern days, also, has it not been at the shrine of death, that the highest efforts of the Michael Angelos, the Canovas, the Thorwaldsens, and the chantrys, have been elicited and exhibited?[42]

Celebrating more than simply good taste, but rather the 'Genius of Memory', proponents of the garden cemetery called upon the theory of associationism to justify eclecticism as well as to assert their own expertise. By choosing works of funerary architecture to include the best examples of all styles and by describing the origins and purposes of ornament and their origins in national practices, cemetery designers illustrated the meaningfulness of objects and species, their forms and appearances. Reformers were likewise able to inculcate an appreciation of such qualities as

simplicity and functionality and disdain for more 'French' ones, such as
'impertinence and affectation'. The garden cemetery could serve as the
sepulchre of a democratic society exhibiting values shared by all civilized
nations. It could also reveal the uniqueness of a given nation such as Britain
by specifying the ethnic and cultural constitution of its people. Strang wrote:

7.5
Contemporary
view of Abney
Park Cemetery,
Stoke
Newington

> If associations too are conceived to be of any importance, as connected with the
> situation of man's last resting-place, as we know they are, more or less, in every
> country, materials for the indulgence of such associations might be here afforded to
> the professors of several shades of religious faith. The Catholic, for example, could
> sleep near a spot associated with the name of the Holy Virgin; the Jew could slumber
> in a cave, like that of Machpelah in the field of Ephron; the Lutheran could place his
> 'dust to dust,' and his 'ashes to ashes,' amid those fair and pure objects of nature,
> with which the followers of that faith love their memory to be associated; the
> unobtrusive Quaker could lie in a quiet and sequestered nook, so encompassed by
> foliage, as to realize, even in death, his love of simplicity; while the strict
> Presbyterian could obtain a grave around the column which proclaims the pure and
> unswerving principles of John Knox![43]

In Loudon's detailed plan for garden cemeteries the horticulturalist and
landscape gardener was obliged to police the excesses of his art. In the
discernment of taste, one would have expected a well-educated Scot to be
versed in the tenets of associationism – so too the teacher in the finer points
of Thomas Gray. The rural labourer or student, however, need not have
understood the subtleties of philosophy or language to benefit from their
intended moral effects and to have appreciated the culture of industry behind

them. The duty of the commoner, rather, was to copy, to recite and emulate. Loudon wrote:

To the local resident poor, uncultivated by reading, the churchyard is their book of history, their biography, their instructor in architecture and sculpture, their model of taste and an important source of moral improvement. Much, therefore, must depend on the manner in which churchyards are laid out, and the state in which they are kept. A country labourer may not have the habits of attention and observation sufficiently developed to derive improvement from the style or taste displayed in the architecture of the church; but there is not one countryman that does not understand the difference between slovenliness and neatness, between taste and no taste, when applied to walks, grass ground, and gardens.[44]

While the 'resident poor' may have gazed in wonder at the varied shapes of marble and porphyry that organized the cemetery, they were not necessarily able to distinguish a sepulchre from a hole in the ground. The landscape gardener, along with the popular cemetery guidebook and encyclopaedia of gardening, was to make such distinctions clear.

The concern for clarity, the abhorrence of 'too many walks ... too many seats and buildings' meant that the cemetery designer was responsible for orchestrating scenographically, audially and even using the sense of smell a particular experience of solitude. The garden cemeteries were perhaps the most regulated of nineteenth-century spaces. In this sense, their popular description as 'cities of the dead' gains true currency. Emerging from sanitary laws and building codes governing tomb dimensions, path widths, drainage, fencings and funerary detail was an enhancement of spatial individuation as evident in the cemetery as in *The Gentleman's House*. A coordinated spatial experience in the former was commensurate with its overall surveillance and control. Strang noted that:

while full scope should be given to the purchasers of ground to express individual feeling by every diversity of funeral monument, it should, at the same time, be understood, that no erection will be permitted that is calculated to injure the picturesque appearance of the garden ... That no lot of ground should be sold of less than sixteen feet broad, in order to preserve, at all times, the appearance of verdure and vegetation over the whole garden ... That while individuals should be permitted to plant any flowers or shrubs over the grave or by the tomb of their relatives, the whole cemetery should be placed under the constant surveillance of a tasteful gardener.[45]

A 'tasteful gardener' himself, Loudon believed that too many trees gave the cemetery the appearance of a pleasure ground. This was at odds with the solemnity of such a landscape just as 'promiscuous planting' was anathema for the systemic botanist. Forming trees into belts and clumps prohibited the free movement of air and the drying effect of the sun, both essential to countering the effects of putrefaction, as opposed to arranging species more sparsely with prevailing winds in mind (see Figure 7.6). Besides remedying such defects, pedestrian movement would be enhanced by locating trees along paths where a foreground was established for distant scenery.[46] Trees might be selected for several reasons. Where their suitability for providing shade or encouraging ventilation coincided with significant historical

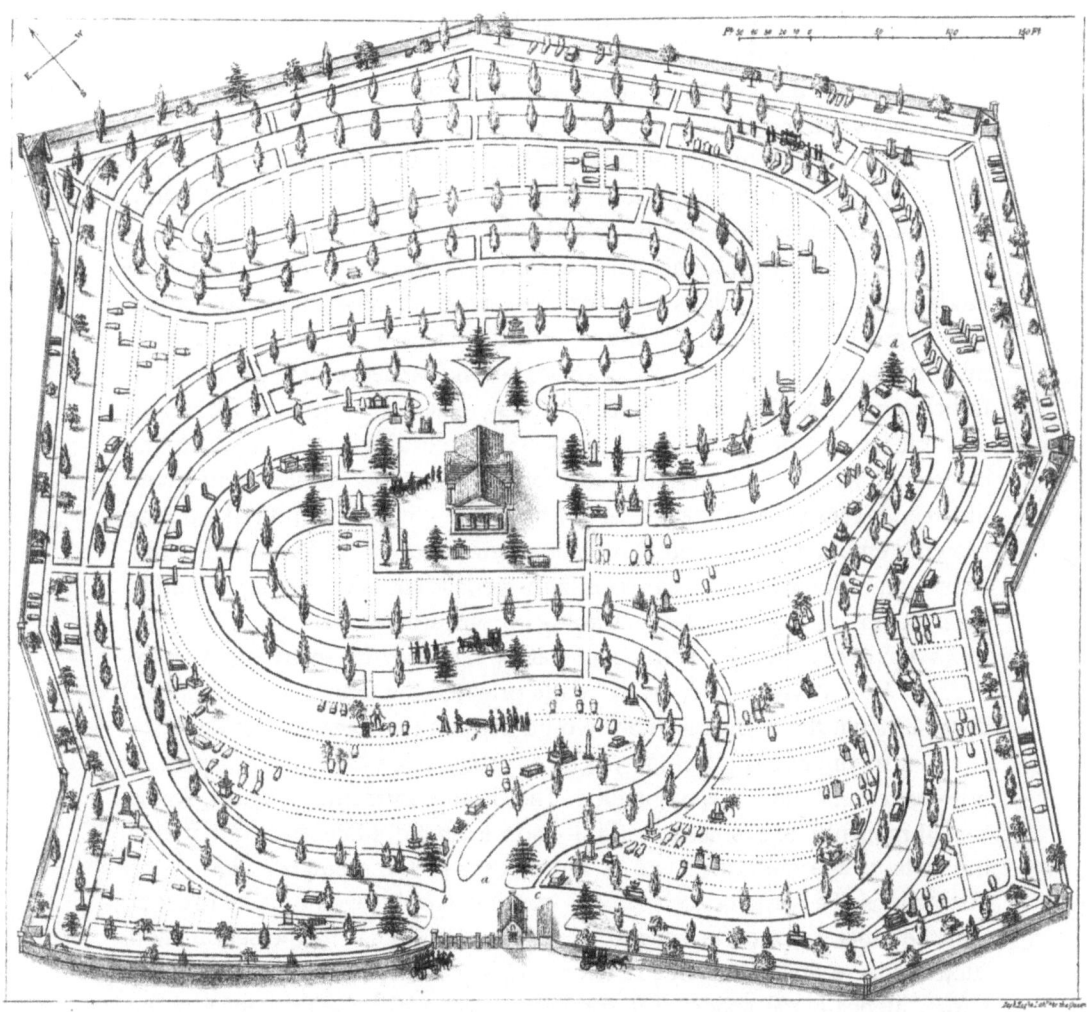

associations, they were celebrated. The common spruce, the reader was told, was 'the cemetery tree of Sweden and Norway', while *Pinus Sylvestris*, or Scots pine, was 'the tree of death and mourning in Russia ... Hence, those streets of Petersburg through which funerals frequently pass are almost always covered with this sign of mourning.'[47]

Such arrangements and associations enveloped or 'immersed' the spectator in a carefully contrived landscape that was, in effect, the image of productive society. More than merely a representation of social coherence, however, the thoughtful arrangement of the constituent parts of the cemetery made it a 'school of instruction' in practices of social cohesion and cultural identity. These elements included objects of 'architecture, sculpture, landscape gardening, arboriculture, botany' and, particularly, the less obvious but all- 'important parts of general gardening, neatness, order, and high keeping'.[48] In this, its pedagogic character, the garden cemetery was far

7.6 'Design for Laying Out and Planting a Cemetery on Hilly Ground', in John Claudius Loudon, *On the Laying Out, Planting, and Managing of Cemeteries; and on the Improvement of Churchyards* (1843)

removed from the eighteenth-century *Bois des Tombeaux* or Romantic woodland interspersed with ruined monument, which poets like Thomas Gray loved to frequent.

To preserve the clarity of designs and isolate elements conducive to reflection, while positioning none to the point of distraction, the condition of fences and herbaceous borders was particularly important. Voicing an old and common belief in the poisons lurking in the soils of traditional burial grounds, Loudon considered the absence of weeds between their gravestones to be a sign of unhealthiness. Paradoxically, though all too common in modern cemeteries, he lamented that they were equally undesirable there. They were, he concluded, 'saturated with the gases of … putrefaction [and] must be productive of malaria'. Far worse, the presence of weeds was an indication of 'negligence and slovenliness, instead of [their] setting an example of neatness, care, and respect'.[49] Accordingly, the 'visit to the cemetery', which Philippe Aries considered a unique development given nineteenth-century attitudes to death, coincides with Curl's remarks on the growing emphasis on memory and commemoration of the past at the time. Both were reinforced by such seemingly pedestrian activities as trimming grass, removing weeds and the cleaning of tombstones.

As for the cemetery's walks and paths, care was taken to avoid sharp borders. Whereas too many walks made orienteering among the exemplary dead difficult, the correct number would present them in the best light.[50] A circuitous panopticism was at work, whereby unease at the prospect of unexpected encounters was instilled. The desired outcome, in effect, was a sort of mournful anaesthesia, carefully administered, of course, so that, like the walls of the monastic's cell or the privacy of the aesthete's library, the distractions of a world conceived of as 'outside' would be dampened in order that the inner voice of conscience might be attended. Thus the sentimental subject was to give free range to his imaginative impulses. There, as Strang wrote, 'Beneath the shade of a spreading tree, amid the fragrance of the balmy flower, surrounded on every hand with the noble works of Art, the imagination is robbed of its gloomy horrors – the wildest fancy is freed from its debasing fears.'[51]

Though the sentiment compelling this voice to speak may have been universal, its language was certainly not. In condemning the British for their disregard for their final resting place at home, Strang quoted a popular travel writer who noted the attention lavished on their cemeteries abroad at Leghorn, Nice and Rome:

Of the Protestant cemetery of Rome, situated near the tomb of Caius Cestius, *Rogers* thus speaks in his *Italy*, 'It is a quiet and sheltered nook, covered in the winter with violets, and the pyramid that overshadows it gives it a classical and singularly solemn air. You feel an interest there, a sympathy you were not prepared for. You are yourself in a foreign land, and they, for the most part, your countrymen. They call upon you in your mother tongue – in English – in words unknown to a native, known only to yourselves.[52]

Such sympathy was the effect of Smiles's indomitable English spirit, though its medium was less the product of religious antipathy and more a sensitivity to cultural difference.[53] In a similar invocation of its inner voice the spokesman for the Secession church at the opening of Abney Park happily allowed himself to be upstaged by the more recumbent members of his audience: 'The pulpit has its voice; but that of the tomb is more thrilling, more impressive [by] far.'[54]

The simplicity in language and sentiment in the epitaph advocated by Conformist and Nonconformist alike during the Victorian era served not only sobriety but communication, as one sees a kind of literary criticism incorporated into the forms of sepulchre, monuments and their inscriptions. Biography and verse combine to engage the readers and to test their astuteness. In 1841 Jane Elliot requested advice from her brother, William Somerville, regarding the choice of inscription in memory of her late husband: 'To be read it should be short – and to claim respect for the living as well as the lamented it should not be as these monumental inscriptions often are – gross thou' [that is, 'though'] misplaced flattery.'[55] Likewise justified was bareness of the sepulchre, as the drapery-clad urn and melancholic angel formed part of a grammar of sentiment. As the broken column represented a life cut short, the sculpted book (faith) and the engraved heart (love and devotion) were symbols commonly offered by guidebooks, as were their botanical counterparts in the lily (purity), the palm (peace) and the willow (grief and mourning).[56] A reading of Thomas Gray was not essential to the appreciation of the familiar forms and flora of sentimentality.

Likewise, in the construction of tombs and gravestones 'trivial or vulgar' forms were eschewed. Writing of the Glasgow Necropolis in 1841, Loudon was to commend its tidyness, for its monuments conveyed 'the dignified idea of being built, and had not the mean appearance of being thrust in like slates, or laid down like pavement', as was all too often the case in English cemeteries.[57] Perhaps the neo-Egyptian entrance to Abney Park resulted from a general fascination with the archaeological trophies of Napoleonic campaigns, as some writers have suggested. Perhaps it suited the taste for simplicity and economy of its engineer designer, William Hoskings. The gates were perfectly suitable for the new use to which they were put, being sober, somewhat two-dimensional and with few competing applications in the realm of architectural or building compositions which would have thwarted their association with death and mourning.[58] It is interesting to note that Hoskings was appointed Engineer to the Birmingham, Bristol and Thames Junction Railway in 1834. In 1836 Isambard Kingdom Brunel had completed his Clifton Suspension Bridge at Bristol. This brings to mind that Brunel's original designs called for the supporting pylons of this structure to likewise be executed in the Egyptian style. Perhaps it was the newness of both public cemetery gate and suspension bridge that called for a simplicity and a style without precedent.

Fears of forgetting

As the nineteenth century drew to a close, the garden cemeteries built in the first half of the century grew less popular. Cemetery companies fell into bankruptcy while, generally, changing attitudes to death in the twentieth century caused many to view with distaste the public displays of mourning associated with the Victorian era.[59] Though attitudes to death may have changed, the moral weight of the past and the civic ideal of the cemetery espoused by reformers in the 1830s and 1840s are commonly promoted by cemetery preservationists today. Mirroring the thoughtful activity of weeding the common grave, the halting of 'tidal waves of ivy' and removal of 'convolvulus and bramble' as well as the self-seeding sycamores and the advancing ash trees which plague cemeteries like Abney Park today have assumed great urgency (see Figure 7.7).[60] In this more recent exercise of 'neatness, care, and respect' can be seen a familiar means of self-reflection and moral improvement in the administration of the dead and the staging of the past.

The garden cemetery was but one of several spaces in nineteenth-century Britain where awareness of the passing of time and both great and humble deeds were cultivated. The museum and public library are two of the more obvious instances. In the home everyday activities organized by the clock, work and social calendars were further orchestrated by lives divided in stages by birth and growth, maturity and old age. These divisions were obviously not new, though increasing life spans, studies of adolescent behaviour and the psychological basis of personality gave them new significance. They were marked by the rooms and landscaped spaces where significant events, rites of passage and memorable encounters occurred. Each moment and place was potentially meaningful, some more than others. If life ended in the grave and was measured by the rigours of everyday life in the home, it began in the unique spaces of the nursery and playground, where the idea of childhood inspired a zone of innocence from which grew a fully formed person.[61] The nuclear family, so much a part of Victorian life, was equipped with such spaces for physical and imaginative play. They formed a 'cordon sanitaire', kept apart from the knowledge of change and death afflicting adulthood.[62]

In the private garden, of whatever extent, appreciation of the past could be reinforced as easily as though it too were a cemetery. Edward Kemp observed in 1850 that:

A tree or plant which we, our relatives, or some known and noted personage have planted, or reared, or tended; a summer-house that is rich in family or other ancient records, or in which we or those we love have thought, or studied, or felt much; a retired nook or secluded little garden, which the fair hands of the departed have, by their former ministrations, hallowed and rendered sacred; may all be abundantly fraught with the beauty of association.

By this benignant law man is linked at once to the material and the spiritual world; and the elements of a garden become pregnant with both poetry and history. The

7.7 Funerary monument, Abney Park Cemetery, Stoke Newington

chords of the human heart are strung responsively to a variety of objects; and a sight, or a sound, or a scent, may at any moment waken their melody.[63]

Such appeals to associations are interesting partly as expressions by philosophers and aesthetes; moreover, they are evidence for a widely held fascination with memory and the passing of time, and they also belie a great fear: the fear which was given voice by the English and Scottish proponents of the garden cemetery and echoed by advocates of rural cemeteries in America. Whom was Loudon addressing when he described the salutary effects of the churchyard, if not a being who *should* remember, though was perhaps unable to? It is through this call to memory that works of architecture and landscape gardening played an important role in the definition of a people, their sentiments and concerns for their fellow citizens. In the gap between the 'material and the spiritual' worlds or between the concept of the social and the idea of the soul, the built environment has emerged as a medium for accommodating one to the other.

As the counter to normalized cognition, 'dissociation' became an increasing problem as the nineteenth century unfolded. As science sought to define memory in terms of its role as guarantor of personal identity, its earlier manifestations appear as such curious mental states as amnesia, hysteria – with its attendant symptoms of blindness and paralysis – fugue, multiple personality – or as it was called by British doctors, doubled consciousness – and, of course, melancholia.[64] It was later in the century that dissociation came to be viewed as an adaptive response to overwhelming trauma.

The study of dissociation appeared in the guise of pastoral care transformed in light of the withdrawal of the consolatory role and moral authority of the Church. It was acted out in the context of self-understanding, the family and the State. It arose out of a generally widespread, though multivalent, fear of disorder. It was provoked by the deleterious effects of urbanization and population increase, of the consequences of movements beyond the familiar church commons or field and of diversifying means of production and of the commodities they produced. With hindsight, efforts to derive associations from particular landscapes, botanical species and architectural objects in the garden cemetery can be seen as various attempts to know and manipulate memory as a key to unlocking an 'inner' world of motivations, desires and sentiments unique to humankind and certainly to each Briton.

Conclusion
Our (dys)functional environment

Concern for time in the nineteenth century was prefigured late in the eighteenth with thoughts on the idea of history. In 1784 Immanuel Kant wrote an essay in which he saw the unfolding of human actions in the world to be as predictable, if not as obvious, as the facts of life and death. Though philosophers might have disagreed about the source of the mind's independence, for Kant, manifestations of the freedom of will were as much subject to universal laws as the 'oscillations of the weather and biological and other natural phenomena'. It was the 'province of History'

to narrate these manifestations; and, let their causes be ever so secret, we know that History, simply by taking its station at a distance and contemplating the agency of human will upon a large scale, aims at unfolding to our view a regular stream of tendency in the great succession of events so that the very same course of incidents which, taken separately and individually, would have seemed perplexed, incoherent, and lawless, yet viewed in their connection and as the actions of the human *species* and not of independent beings, never fail to discover a steady and continuous, though slow, development of certain great dispositions in our nature.[1]

Over the course of the following century, the human species was studied in various ways, with biological or life sciences and statistical analyses imparting new meanings to events characterizing life, its duration, rhythms and patterns of growth, inheritance and development. Today, these have been complemented by broadly based environmental studies, ecology and evolutionary science. Most studies reinforce the view implicit in the above quotation on the distinctiveness of human experience of the natural world. Kant's universal history was human history, after all. In previous eras human uniqueness was a matter of human superiority and mastery over other species. It raised the prospect of the wholesale transformation of natural resources. Today, thoughts are more likely to be about the threat of human activity to the environment that sustains all species.

The life sciences encouraged a sense of the uniform unfolding of events in the Victorian home by providing explanations for why one action caused another to occur. They imparted a degree of novelty, mystery and suspense to the myriad of organic operations occurring there. Residents were led to believe that consequences, whether 'good' or 'bad', invariably followed each moment of life, if only they waited long enough or looked for them closely enough. In response, writers encouraged residents to impose patterns of order in the house and garden to stimulate the mind by arranging furnishings for multiple activities and plant species so that the seasons might pass with their own 'peculiar beauty and charm'. Some, like Robert Kerr, encouraged readers to rectify deficiencies in the planning and comfort of houses large and small by imagining how these might actually be experienced. Residents were encouraged to imagine how houses might facilitate movement or augment the body's recuperative abilities or be positioned so that their windows might enhance the varieties of *chiaroscuro* choreographed by the sun's daily transit. A sense of the past and future led social reformers, landscape gardeners and architectural critics in the 1830s and 1840s to devise new forms of interment, and Edward Kemp to encourage householders of great and limited property to value all things great and small, planted, reared or tended. These varied efforts were played out on the moral landscape of Britain as well as in its living spaces by writers of varied interests.

The cultivation of this landscape was a goal shared by scientists, reformers and statisticians when asserting that some individuals were not easily improved. They seemed to veer from norms of reason and behaviour and standards of perception – 'art' or 'taste' in terms more familiar to the architectural critic or landscape gardener. These norms became apparent when one observed society as a whole. In 1869 Francis Galton published his celebrated work, *Hereditary Genius*, making variations of character an object of study by analysing the distribution of mental attributes among urban populations.[2] The circumstances amenable to solitude and repose that reformers sought to create in the garden cemetery, or architects in a gentleman's study, were among a number of conditions lacking in the slum dwellings of the poor. Their absence was thought to be a cause of degeneracy. Galton was responsible for studying variations within the human species from a quasi-evolutionary perspective. It was a viewpoint where character, like an individual's physical attributes, their height or weight, were the consequence of a specific constitution and a history of adaptations, for better or worse, to specific environs.[3] Whereas Darwin evoked evolutionary time-frames measured in millions of years to explain the forms of living species and their habits of life, Galton relied on a more compressed scale. He defined the cause of racial and class superiority in abstract, biological terms. Among these, 'stock' was long familiar to breeders of plants and horses and readers of books on English heraldry and peerage, such as those by the nineteenth-century genealogist, John Debrett. The term initially denoted lines of descent, rather than particular characteristics of an individual per se. For Galton and his followers, however, stock became a measure of intellectual ability, which was itself the surest index of the state of all mental and physical faculties. It became 'the measure of vital energy itself'.[4]

The notion of racial superiority has since been scientifically repudiated.

However, the turn to biology and related disciplines as a source of explanations for a range of phenomena in nineteenth-century thought has had significant ramifications for the history of building forms and landscapes. Viewing human history through the lens of life science, there appeared over the course of the Victorian period two broad categories of concern addressing more or less the question as to exactly *where* the vital energy that coursed through all living beings drew its charge. Exactly where, in the case of humankind, lay the mind's independence and distinctiveness? This question concerns us today. One source of possible answers lies in a world of environmental influences broadly conceived, both natural and social. The second concerns an inner world of motivations, desires and sentiments unique to the human species. Played out in the landscape of the Victorian house and garden and through various theoretical and practical means, garden features and spaces exemplifying conditions of enclosure and exposure, proximity and distance, interconnectivity and so forth became means for thinking of human beings as somehow situated between these two worlds. They were means of inhabiting the one physically and occupying the other in the mind. Designing these landscapes and interiors became a means of manoeuvring between the two. It required – and continues to require – a careful balancing act.

Landscapes of the mind

In the environs of the garden cemetery it was hoped that through the cultivation of solitude along winding paths and spaces made distinct by carefully placed monuments, the rigours of life would be assuaged. It was intended that the vital energy of an enterprising British nation should be restored by cultivating the sentiments of its people, their feelings of charity, common humanity and civic pride. In the reformed asylums of the 'mental hygienists', on the other hand, solitude was sought through straight corridors and dark and quiet rooms devoid of distracting clutter. Their purpose was to counter the propensity of the insane to form mental associations wildly and uncontrollably. Opposing the image of the contemplative in the garden with the lunatic in the asylum recalls to mind the wild man, a figure that likewise had difficulty in associating ideas and things and was obliged at times to learn how. Separating these figures historically was a focus on the 'psychological' dimension of human experience that arose in the late nineteenth century.

Galton presumed that confined spaces and limited sensations could restore the mind to health just as Etienne Condillac had previously believed in 1746 that the deprivation of any one sense could affect mental capacity.[5] The French philosopher's work inspired subsequent efforts to educate primitives and feral children by introducing them to a range of objects and experiences. By acquiring an ability to discriminate between them, complex ideas and values would follow.[6] Impaired by physical disability rather than historical circumstance, as feral children were, the deaf and blind were taught sign language and other techniques devised to bridge the gap between stimulus and sensation. A new object of scientific investigation grew out of Condillac's 'sensationalist' theories. As the philosopher construed the relation of mind to reality in terms of representations presented to understanding through the mediation of senses, of which vision was an ideal model for all senses, a 'surface' formed upon which it became possible to *think* about human psychology.[7]

John Locke argued in the Introduction to his *Essay Concerning Human Understanding* that an inner faculty of reflection formed ideas that mediated between the mind and phenomenal reality. These ideas were derived from sensations and experience of 'visible resemblances' or 'things without'. If ideas were a form of inner vision, words were secondary, arbitrary marks upon the world. They were derived by humankind to facilitate communication. Condillac extended Locke's theory, reversing the order of relation between ideas and signs. Representations served to reveal the method of the mind's operation. Ideas were only capable of serving a process of understanding, and therefore of being combined to form more complex ideas and concepts, once they were incorporated into signs. As a system of signs, language was the means to, not the end product of, reason. One consequence of this reversal was that an ability to read or interpret the function of a thing or the order of a landscape or room – each being signs of a kind – was tantamount to valuations of intellectual ability or mental character. To understand the purpose of an enclosure as providing necessary shelter or the form of a building or plant species as 'natural' or 'artificial' or as affording visual pleasure or interest was tantamount to valuations of other sorts. They allowed one to decide whether or not the one (shelter) was *genuinely* comfortable, say, and the other (building or plant species) *truly* natural or interesting. In abstract terms, understanding and thinking about building forms and landscapes requires concepts and language to express them. People do not know what they are thinking until they 'say' it to themselves. Nineteenth-century writers like Cuthbert Johnson, Robert Kerr and John Strang and most clearly John Loudon are notable for having 'taught' their readers 'how' to think, meaning in a certain way. This involved a language of forms and purposes, of the 'fitness or propriety' of constructions, or the 'utility of the end' that buildings and gardens were intended to serve. It commonly required commentary on or illustrations of 'contractedness' or 'wasted space' in a house or garden and broad moral pronouncements that one should not mimic nature or, in one specific case, that a 'clumsy mound' in the cemetery was not enough to remind the British people of the need for virtue.

The widely promoted goal to balance diversity in form and moderation in composing the landscape of the Victorian house and garden, like the garden cemetery, imposed a unique burden upon the inhabitants of these spaces. It required a degree of critical acuity of the resident or pedestrian. It obliged them to police not only their own behaviour and that of other family members or visitors, but their own tastes and perceptions, their thoughts and sentiments as well. Notwithstanding the deficiencies of planning in the home, as applied to the ostentatious gardens of the French, according to Loudon, these spaces offered important opportunities for individual cultivation. By observing, in his words, 'the exaggeration of either beauties or deformities, the causes of the pleasure or the dislike that they excite are more easily discovered; and consequently our resources for enlarging the one or diminishing the other increased'.[8] The expansion of the mind through enhanced association lies at the heart of the Victorian home's popularity as a site for moral improvement. It accounts for the ready extension of associationism, as either a matter of philosophical interest on the part of aesthetes or, expressed in everyday language, as part of the 'surface' for exploring a psychological response to landscapes of various kinds.

The preceding comments raise an important point of methodology concerning relations between theory and practice, history and present-day concerns. Thoughts on the environment, like those on human psychology, did not appear out of thin air. Robert Kerr's treatise on residential design, for instance, relied heavily on drawings and descriptions of aristocratic palaces from former times for the exemplification of key principles. While there was certainly enough material in the book to build a better palace and complete the catalogue of rooms commensurate with their occupants' high social status, readership was not limited to the noble or wealthy alone. Rather, the treatise was one of many that allowed discussion of 'noble' principles – those characterized by ingenuity and skill or a distinctiveness and integrity of character – formerly associated with one class of people or one type of building to be transposed onto other kinds of people and other types of buildings. The book was a textual surface from which new thoughts on sound planning, comfort and the character of the inhabitant could spring. In common works of nineteenth-century domestic architecture and household economy, the characterization of figures in rooms and passageways or landscapes, though hardly novel in itself, engaged with other forms of literature and relatively new discourses, such as psychology and sociology, anthropology and ethnography. These made it a common assumption that the figures reacted to things of deep significance in their surroundings.[9]

Thoughts on human psychology in the modern sense of the word depended partly on developments in theory, but equally important were practices and governmental initiatives whereby the results of investigations into mental character and ultimately morality assumed relevance and value. For Rousseau and his contemporaries the study of primitives primarily served philosophical or rhetorical purposes, providing material for speculating on human nature. The training of feral children and the development of new means of communication for the deaf in his day helped answer perplexing questions regarding the uniqueness of humankind based on its use of language. Primitives, wild boys and the deaf and blind were curious creatures in that they were outcasts from society. However, the psychology of the visitor to the botanical garden or resident in the suburban villa, the gentleman in his study or homemaker in her kitchen, on the other hand, became a matter of importance for the structure of society as a whole. Predicting sentiments and behaviour was tantamount to altering, correcting or improving behaviour.

The instinctual drives that in one way or another came to characterize the vitality of life served in yet another to be its undoing, imparting a degree of indeterminacy into human relations. Uncontrolled self-reflection led to narcissism, misdirected desire to auto-eroticism and degeneracy. Rendered by Freud into so many complexes, these too found their expression and containment indoors and behind garden walls. The home imparted an organic structure to family life, providing a site for the actualization and management of vital processes and imposing dominion on the individuals accommodated within. The home became an organic entity like the living organism itself that 'weaves back into the unique fabric of its sovereignty both the visible and the invisible' and serves as the measure of life and lifestyles.[10]

Environmental concerns

The fear of mental dissipation and the concern for its consequences for national vitality, as expressed by the Victorians, was linked to the second category of concerns and anxiety manifest at about the same time. It was the apprehension concerning degraded environs and their effect on well-being, physical and mental health. The belief that one's surroundings could aggravate an inherited capacity for disease coincided with mounting evidence that the transformation of nature by humankind was occurring with such magnitude as to threaten the basis of nature's laws and economies that ultimately was the basis for the well-being of all species. Britain's industrial landscape began to provide the ground for a raft of such concerns in the eighteenth and early nineteenth centuries. From the 1860s onwards, however, the anxieties of many regarding the crime-ridden slums and smoke-filled skies were extended to include a broader picture of deprivation.

Reactions to forms of environmental degradation were evident in the masterful narratives of reform like George Perkins Marsh's *Man and Nature* or Richard Jefferies's *After London*, but also in writings which have since become more obscure. For instance, a subtle but passionate plea for what, but for its association with picturesque sentiments, would be termed 'ecological diversity' was implied in the now seemingly quaint or nostalgic objections of some Victorian gardeners to the disappearance of familiar and much-loved flowering plants. It is not only approximate publication dates that draw together these various expressions of environmental concern with Galton's *Hereditary Genius*. Rather, it is that each, either more or less explicitly, dealt with specific living beings as subject populations, exhibiting a certain vitality or constitution, and which in turn were affected by their living and non-living surroundings – environs in which humankind played a key role. Preferences for either geometrical or English flower-beds by nineteenth-century landscape gardeners and their clients coexisted with criticisms that horticultural systems of whatever style could prove lifeless if infested with overly bred and grafted plant species bereft of vigour. For some the availability of flowers in wide-ranging colours but without scent, or of trees that fruited out of season although proven to be sterile, represented the denial of nature. This was because the human desire for variety and fashion took precedence over a key function of any organism, namely its capacity to reproduce itself. For plants this was served by flowers that attracted insects for pollination and producing seed. Through forms of dualistic reason and an emphasis on language as a measure of humankind's privileged status, the categories of nature and culture and the urban and the wild were distinguished from one another. So, too, the term 'weeds' was distinguished first from 'fields', then from 'natives' or indigenous plants notable for having been unaltered by human interference. In the absence of foraging animals and the improving hand of humankind, even common species could become weeds in Jefferies's ecological horror story.

The image in *After London* of once lovingly tended fields grown wild through neglect was mirrored by the horror of nature too closely controlled. Many nineteenth-century writers expressed nostalgia for 'old-fashioned' gardens. For the Victorians, as Michael Waters observed, the qualities of variety, simplicity, harmony and unforced abundance associated with an older style of gardening coincided with qualities of life commonly associated with pre-industrial England.[11] The appeal to traditional horticulture often carried with it a critique of modern, scientific efficiency and rational improvement and an explicit rejection of aspects of the modern style of landscape gardening: the separation of flower- and kitchen-garden, the monotony of bedding out, the reliance on monochromatic flower schemes and the loss of traditional garden plants. From about the same time as the publication of *Man and Nature*, garden writers were investing cottage gardens with preservationist intentions.[12] These tendencies persist today in wide-ranging projects. From a cultural perspective they include efforts to preserve or restore period gardens for what they tell us about the shaping of nature in different eras. From an ecological viewpoint they include calls for the safeguarding of remnants of native forest and peat bogs in Britain, efforts to find sources of timber other than tropical rainforests and alternatives to the mountains of peat moss consumed by gardeners, without destroying reserves elsewhere. If many ecologists had their way, the American grey squirrels would be sent packing from British parks and gardens.

Darwin at home

Counterpoising the nineteenth-century glasshouse with the Victorian house and garden reveals two important divisions within the landscape of mind and environment that we have inherited from this earlier time: the division between the home and its broader urban context. The fast pace of change in Britain's towns and cities in the nineteenth century mirrored the rate of geomorphic change in many places in the natural world, most obviously its tropical regions that glasshouses sought to reproduce in miniature. To be observed, both were given boundaries so that extraordinary forces could be experienced in a heightened sense. On the tropical island, change was evinced by a superabundance of living species, particularly plants and seemingly endless cycles of exuberant growth and decay. As the fundamental unit of the city, the home served to reproduce broader urban and industrial phenomena. It benefited from developments in sanitation, heating and the mass production of consumer goods, but also harboured deadly miasmas, toxins and rising levels of indebtedness. It was an analogue of nature, a reserve against the fast-encroaching city. The home was also the vanguard for the control of nature, its confinement and enclosure.

George Levine has identified in *Darwin and the Novelists: Pattern of Science in Victorian Fiction* certain recurrent themes or motifs in Victorian literature that suggest the pervasiveness of biological science, particularly evolutionary biology, in late nineteenth-century thought in Britain.[13] These themes underscore the appropriation of nature by the theologian and evolutionist as well as the novelist and householder, the homemaker and domestic gardener, and also highlight concerns for the environment at the time and subjective responses to it. They form a framework for understanding the 'dangers of imprudence', as Marsh described the consequences of interfering with the spontaneous arrangements of nature.

These themes include, first, the relevance of scientific certainty and the 'rule of law' for most aspects of everyday experience. In the domestic sphere, the garden was shaped by advances in agricultural chemistry, while it became possible to organize daily activities, mindful of the operations of organic and inorganic nature. As *Cassell's Household Guide* advises:

Chemical operations are performed every hour in the day in every household. From the moment when the housemaid strikes the first match in the morning to the moment when the last candle is extinguished at night the forces of chemistry are at work; and even when all is still and the gentle breathing of the sleeping inmates is the only perceptible movement in the house, that very breathing involves a beautiful and complex chemical process.[14]

Policing the boundaries between earth, air and water, in sweeping, stopping draughts, drawing water or washing down surfaces in the Victorian home not only bestowed an 'imperceptible, but palpable comfort', but also gave useful occupation to the mind. These activities filled 'those odd moments of time that are too often listlessly dawdled or idly gossiped away', encouraging the 'constant round of useful employment that keeps the mind cheerful, and thus helps materially the health of the body'.[15] Given the homemakers' efforts to account for the organic processes occurring in the garden or the belief that household tasks were best performed when organized efficiently with minimal waste of energy, these activities acquired new significance. Focus on the 'details of the here and now' became sharper. Easily overlooked natural phenomena and domestic activities were worthy of study by being part of a larger system of order and the values it entailed. In literature, this sharpened focus on the everyday has been associated with the theoretical aims of realism. In the house and garden, the novelist's effort to portray everyday life was paralleled by the householder's and designer's attempts to accommodate it in ever-greater, more 'realistic' or 'natural' detail.

The prospect of environmental mastery in the home raises a second and familiar theme, namely a fascination with nature's wholeness or plenitude. For the natural theologian, abundance was a sign of God's beneficence in having spread creation everywhere. For Darwin and his followers it promised 'overcrowding, waste and disorder'. It portended a reality of randomness and chance. For the gardener, whether spiritually inclined or scientifically minded, the abundance of nature was something to be emulated. It inspired what Robert Kerr advocated in 1864 as a 'carpet of design' to surround the house and make it 'fit' into nature. Abundance was associated with virtue. It was a sign of respect for a natural world that was nonetheless transformed through human enterprise. The cultivation and enjoyment of abundance were attributes of a burgeoning material culture where carefully arranged landscapes of varied and often valuable botanical species were but extensions of interiors composed of wall and floor coverings, furnishings, curios and other possessions.[16] Homemakers, more likely armed with manuals of practical advice than evolutionist tracts, faced abundance with the same fear of 'overcrowding, waste and disorder'. This was despite, or perhaps because of, best efforts and intentions in ordering their environment.

A third theme, adaptation, was capable of inspiring contradictory explanations for the abundance and diversity of creation. For the scientist, adaptation emphasized relations between living beings or, in other words, just how things fit into the overall picture. In the domestic sphere, the underlying principle of adaptation brought home a major rift in the range of attitudes toward nature that came to be espoused by the second half of the nineteenth century. There appeared, on the one hand, a seemingly limitless realm of natural resources and fundamental elements and spaces accommodating the expansion of human vitality. On the other, there were environs – more limited or circumscribed – that ultimately determined human character. In the first instance, the home was infinitely malleable, available to any number of human designs. In the second, the character of individuals – their constitution – could be seen to be largely predetermined given the habitats in which they were born and had lived. The one reflected the other, but a good home could improve character and vice versa, as much as being a sign of it.

The principle of adaptation raises questions of the possibilities *for*, and the limits *of*, human autonomy and freedom and, ultimately, environmental mastery. In the title of his book of household advice of 1892, *Is My House Healthy?*, James Cameron invoked a realm of uncertainty shared by innumerable householders, not only of concern to 'sanitarians'. By the end of the century, the author's observations were paralleled by a commonly held truism, a new take on the nature and culture divide, whereby:

Conflicting schools of thought assure us, on the one hand, that 'Man is man and master of his fate'; and on the other, that he is the creation of his environment. To whichever side the dialectic victory may in the long run incline, one thing is clear to the sanitarian – that so far as that important element in the environment of a family the house is concerned, the householder himself is to a great extent master of the fate of those beneath his roof, the 'hostages he has given to fortune'.[17]

Today, of course, it is clear that the earth's resources are not unlimited or, if they appear to be, this is so for only the small, richest portion of the earth's population. Like the Victorian homeowner's, our houses and gardens remain bounded by walls and routines of care. Our attention is directed alternately both outwards and inwards, though the former, broader environment seems to exert an even stronger pull than in years past given *our* awareness of disturbing 'atmospheric phenomena', which the Victorians never dreamt of, such as global warming and the greenhouse effect.

A fourth theme that made the Victorian home a laboratory for exploring scientific ideas is perhaps the most fundamental and one which most clearly links that previous era with our own. It incorporates the previous three of Levine's motifs and directs us back to the idea of Loudon's artificial climates. It is ecology, evidenced by the sensitivity one finds in works of nineteenth-century literature and science, domestic economy, gardening and design to the 'interrelationships between living organisms and their surroundings', the forms of economy between them and the sense of balance that sustains them. Adaptation brings to ecology an acute awareness of time as essential for any living being's evolution to suit their surroundings.[18]

Successful less as a testing ground for ultimately proving one idea or school of thought or another, but rather as a domain for asserting scientific certainty and human values, the Victorian home was a place where even the lowliest of creatures and most obscure organic phenomena could be informative. They imparted lessons of both the complexities of the organic world and its vulnerability. Lawns, for instance, with their carpet of turf 'so peculiar, and so dear to our rural life', were made possible by the incessant care of the gardener. However, it was the lowly earthworm and other, unseen organic forces which rendered the soil fertile that remained little understood. Though the absence of 'all powers of perception' in such operations was a foregone conclusion, the organic world nonetheless provided living models of endless effort and adaptability. For Johnson they 'add another proof to the endless evidence we possess that nothing was created in vain. As I have elsewhere had occasion to remark, almost every advance in natural philosophy seems to increase our acquaintance with the creatures that are our fellow-labourers in the culture of the earth.'[19]

Darwin himself had argued as much, writing 'The Formation of Vegetable Mould', which was published in 1881. The essay proved instrumental in directing farming and gardening towards organic principles by reinforcing the dependence of soil formation on living organisms. His in-depth study of the actions of earthworms became popular some years after its initial publication, due, paradoxically, to the persistent influence of chemical methods on agricultural science.[20] Until Darwin's essay appeared, no one had recorded how earthworms were responsible for exposing surface soil to the air and for sifting and mixing soil particles through digestion. Studies of the effects of chemical agents on organic matter produced unexpected results, particularly later in the nineteenth and early twentieth centuries. Then, efforts to produce humus on a large scale revealed that the continued used of chemical fertilizers and sprays for keeping the pests of fruit trees at bay had led to the destruction of earthworms on a large scale, with possibly dire consequences for humanity as well.

Darwin's essay showed that other species formed an inextricable part of Kant's human history. His close examination of the habits of worms was responsible for drawing attention to creatures that 'have played a more important part in the history of the world than most persons would at first suppose'. More importantly, it raised awareness of the interdependence of animate and inanimate matter forming life itself.[21] In leaving evidence of their presence on the earth in the form of innumerable castings – in part of England, evidently, some 11 000 kg (10 tons) of dry earth per half hectare (over 1 acre) passes through their bodies – earthworms preserved traces of human habitation as well. Archaeologists, Darwin wrote, 'ought to be grateful to worms, as they protect and preserve for an indefinitely long period every object, not liable to decay, which is dropped on the surface of the land, by burying it beneath their castings. Thus also, many elegant and curious tessellated pavements and other ancient remains have been preserved.'[22] In cities like London, Paris and Rome, however, where the worm found no home for its industry, enormous beds of rubbish many yards thick were found beneath the surface, to mark the course, if not necessarily the progress, of human history and civilization.

The value of adaptation, entailing the broad concern with forms of dependence between plants and animals and their surroundings and, more generally, of change in the natural world, was more or less explicitly addressed in the house and garden. It was in fact the arrangement of the domestic sphere that brought the web of vital processes in which humans found themselves 'home', that is, made them 'real'. Regarding change in both natural and domestic spheres, earlier ways of relating organic phenomena to their underlying causes in the former gave way to the general belief that all phenomena can be explained causally, according to a logical sequence of events. Various interpretations of nature's plan, such as mechanicism, miasmatic theory or natural theology, gave way to a position called 'uniformitarianism'. By the time of Darwin, there appeared an understanding of life involving a dynamic continuum in which, according to Levine, '*everything* is always or potentially changing and nothing can be understood without its history'.[23] Balfour wrote:

To be born – to arrive at maturity – to die and to decay, is the sum of the history of every being that lives, from man in the pomp of royalty and the pride of philosophy, to the gay and the thoughtless insect that glitters for a few hours in the sunbeam, and is seen no more; from the stately oak, the monarch of the forest through successive centuries, to the humble fungus which shoots forth and withers in a day. How simply, yet how expressively, are these changes described in the words of the sacred writer: 'Our life is as a vapour, which appeareth for a little time and then vanisheth away.'[24]

A belief in the contiguity of organic existence drew the attention of the religious and the evolutionist, the householder and the domestic gardener to the divisions between species and to the conclusion that all manner of living beings exist in a state of flux. Time made it possible to think about the members of the human species as having a constitution and individuals an inherent hardiness or a predisposition to illness or decrepitude. In this regard a constitution was a measure of adaptability. Adaptability imparted an even greater significance to one's surroundings than that which nature's fundamental elements would reveal. Surroundings could prove as influential in the past and future as they were in the present. They were the source of numerous organic phenomena, determining one's health and well-being, but also promoting nostalgia and the desire for a world that was comfortable and risk-free. Between the significant moments of buying and selling a house, the risks associated with each were compounded by threats of innumerable, potential domestic disasters.

Living in glasshouses

As for glasshouses, like the Victorian home and its compartments and the garden cemetery, they were places where life and death, the normal and the pathological, were made evident. There, the meanings of biological existence were defined along with the passage of time. These spaces made possible assessments of human identity, character and integrity and assertions of natural and cultural values, where forms and appearances provided material for different kinds of evaluation. They were places where novelty was accommodated, though as previous chapters illustrate, 'newness' was not only an outcome of designs and good intentions, but included revelations of the limits of human understanding, inventiveness and technology as well. They were places where 'knowing' something about nature was akin to 'doing' something about it – improving it, re-arranging it or finding reasons why nature proved resistant to change.

Fifty years after Loudon exposed the 'genial climate' inside the curvilinear forcing-houses to public scrutiny and questioned what they should look like, subsequent developments in iron-and-glass construction and improvements to heating systems made the conservatory a common addition to the middle- and upper-class Victorian home. At the same time, David Thomson was able to warn of the obvious artifice and the potentially injurious effect of these structures on life forms, plants and human beings: 'The practice of horticulture has been regarded as the most healthy employment and most delightful recreation in which human beings can be engaged. This remark holds good of all its branches, unless it be the forcing of flowers and fruits under glass, which is adverse to physical well-being.'[25]

It was not that Thomson disliked forcing-houses, or forcing-beds, forcing-fields or forcing-frames. Rather, by the time he wrote his *Handy Book of the Flower Garden* (1868) nature was considered to be an immense, unknowable and inimitable entity. Lingering enthusiasm for artificial climates was tempered by thoughts that perhaps the environment could not be – or *should* not be – forced to do or be anything at all.

Revealingly, glasshouses of various types and appellation have entered the stock of architectural terms we use to describe ourselves. Taken together, they are hardly unequivocal, suggesting protectiveness and exposure, nurture and fragility alike. While an image of security is found in 'everyman and his castle', sanctuary is lacking for 'people who *live* in glasshouses'. Common parlance allows us to 'hothouse' something if we want to cause its early maturation; while to be a 'hothouse flower' is hardly a commendable epithet, but a condemnation of artifice and the perversion of one's nature. Meanwhile, the 'greenhouse effect' is commonly thought to be a function of deforestation, hydrocarbon emissions or one or the other kind of human negligence – basically, our fault for pushing nature too hard.

It is ironic to consider that the Palm House at Kew is indeed a precursor to much modern architecture, though not for functioning to preserve the living things housed within it, but for killing them; not for encouraging the penetration of life-giving sunshine, but for obscuring the sun; not for recreating the warm humid vapours of the tropics, but for rusting to pieces. In contemporary terms, what is commonly called 'sick building syndrome' encompasses both the dead exotics of the Palm House in 1867 and the ailing office-workers of more recent times, a relationship between all that is normal and abnormal or pathological in life heightened by the technologies that define the one and the other.

Living in a vital landscape

It is tempting to look further for precedents in previous chapters for present-day environmental concerns or rather for a perspective, theoretical framework or 'surface' formed by past ideas and practices that illuminate contemporary ones. It is tempting to discern a common desire for environmental control in the palm houses at Bicton Gardens in Devon and Kew Gardens in Surrey, Loudon's visionary schemes for the Birmingham Botanical Horticultural Garden and, most recently, the Eden Project in Cornwall (see Plate VI). Completed in March 2001 and designed by architect Nicholas Grimshaw and engineer Tony Hunt, the Eden Project comprises a series of large conservatories, biospheres or 'biomes' housing plant species representative of diverse ecosystems and bio-climatic regions. The Humid Tropics Biome alone is the largest conservatory in the world at 240 m (787 ft) long, 110 m (361 ft) wide and 55 m (180 ft) high. Wrought-iron sash bars and glass seem prehistoric compared to the roof of tubular steel geodesic domes and translucent, triple-layered ethylene tetrafluoroethylene covering Grimshaw's design.

Because the dome has surfaces less than parallel 'to the vaulted surface of the heavens', the sun's orbit is a less obvious concern in the Eden Project than it was in Loudon's and Mackenzie's schemes. Rather, a catalogue of high-tech materials and technologies supports its artificial climates. A celebrated 85 000 tonnes of soil was created from china clay and organic waste. A solar-energy system warms the air and stores heat through the night, while recycled water from misters and waterfalls supplies the necessary humidity. A ground-level irrigation system keeps the soil moist so that visitors 'won't have to put up with the rainforest's 1500 mm (60 in) of rain a year!'[26]

It is tempting to wonder whether or not a visitor to the Eden Project in the twenty-first century would discover – like his counterpart in 1846 upon entering the Winter Garden in Regent's Park, built by Decimus Burton and Richard Turner – a veritable 'garden of delight'. In either situation we too might wonder 'Is this winter? Is this England?'[27] Visitors to Cornwall today are likely to have their imaginations stretched even further and find a place that 'looks like another planet' straight out of a James Bond movie.[28] Given the continued popularity of the Palm House at Kew and the care lavished on its restorations, it seems more than 'suitable for horticultural purposes', as Burton originally proposed. It too appears a 'stage where science, art and technology blend to tell the story of our place in nature', just as the Eden Project and its website announce to visitors.

While relatively few species among the 1000 or so housed by the Eden Project so far have suffered the fate of the botanical collection in the Palm House at Kew in 1867, evidence of – and cause for further – human interference in the balance of nature accumulates. A victim of its own success, the project has been condemned by groups of environmentalists for the high levels of pollution afflicting the Cornish countryside as a result of hundreds of thousands of vehicles arriving at the site daily.[29] Whereas visitors to the glasshouse in the Victorian era sought refuge from dull, damp and messy kinds of days, the Eden Project has taken out adverts to encourage tourists to stay away on rainy days to minimize the impact of excessive numbers on overtaxed facilities. Meanwhile, adding to the din of traffic, audio recordings of seagulls crying in distress have been broadcast to stop the birds from pecking and puncturing the project's plastic domes.[30]

While it seems credible that the novel world of the nineteenth-century glasshouse prefigured the Eden Project's biomes in some ways, a more direct comparison carries certain caveats. The terrain surrounding the Palm House at Kew was artfully arranged to present the structure in the best light possible, though it remains, basically, an object on an English lawn with neo-Classical temple, heraldic statuary and a reflecting pond to complete the picture. The biomes of the Eden Project nestle in the crater of an abandoned clay pit, surrounded by the Cornish countryside, but also landscapes typifying regions from countries such as Chile and India. These illustrate crops and agricultural practices found in these places and tell the history of plants that have been used by humankind over the millennia for food, shelter, clothing and medicine.

In the years between George Mackenzie's advice to horticulturists in 1815 and the twenty-first century, biological science, particularly environmental science and ecology, has come to offer a frame of reference for making sense of environs both within and outside the conservatory. There is a concerted effort in the Eden Project to enlist visitors in what its creator called the 'battle for environmental sustainability and biodiversity'. There is an ongoing effort, shared by curators of nineteenth- century botanical gardens and zoos, to enlighten spectators rather than entertain them with just another 'glorified theme park'.[31] Positioned within the different countries and ecosystems imitated in the biomes and the broader environment, amidst the plants species and 'appropriate' animals (insect-eating tree frogs and lizards, mainly including one species of nocturnal gecko to control pests even at night), viewers *become* the attendants – the 'gardeners or curators of the different productions', as Loudon would have called them. Of all the species arranged within, humankind alone is ultimately responsible for sustaining the ecosystems and sense of balance these productions represent, all the while 'habited in their particular costumes'.

Recently, in an entirely different arena, an extension to this framework of biological understanding has come from the field of evolutionary psychology. This development prompts arguments for and against the complete biological determination of *all* aspects of human life, not only the relationships arising between humankind, its culture and its living and non-living surroundings, but even human consciousness itself. Some theorists in architecture and the visual arts and those with an interest in landscape and perception or 'environmental aesthetics' have embraced evolutionary psychology to a greater or lesser degree to explain the basis for 'forms' of human culture, perceptions, aesthetic preferences or tastes.[32] Interpreting Loudon's comments from such a perspective, an extreme position would assert that even particular habits and costumes along with patterns of line and colour, garden bedding systems and arrangements of furniture, needs for shelter and refuge as well as monumental forms are determined by humankind's evolutionary history.

There are many arguments against such a position – too many to detail here.[33] One worth noting is the criticism that such an interpretation of human culture, perceptions and sensibilities in terms of evolutionary psychology has a history itself. In eras prior to our own, it was possible to think otherwise. Perhaps since the era of the alchemist and the age of the machine, 'knowing' something about our own mode of adaptation and past has actually guided, in some small way, the course of our evolutionary development. One, easily overlooked development that suggests this very possibility is evinced by recent studies of the rising frequency of allergies in Western industrialized societies. By way of a brief illustration, consider the chain of events that followed the growing realization in nineteenth-century Britain of the importance of sanitation in the home. By knowing about the presence of miasmas and then germs and by forming hypotheses to explain their consequences for health, the Victorians adopted extensive measures to remove the possible sources of contamination from house and garden. Since then, as scientists have concluded, our homes have become *too* clean, our water *too* pure and – calling to mind a contributing factor to the sick-building syndrome – our air *too* disconnected from the outdoors.[34] As a result our bodies have evolved to respond accordingly and lost certain immunities to allergens, the organic and inorganic toxins our ancestors once had, even though, on the whole, we are living longer – in other ways, 'better' – than they did.

Returning to consider further the terms of reference that connect eighteenth-century natural history with nineteenth-century biology, environmental and evolutionary science, there was extensive commentary in the nineteenth century on the form and appearance of certain classes of buildings and landscapes, such as glasshouses or wild gardens. There was much thought given to designs for specific types of inhabitants such as Kerr's English gentleman or Loudon, the invalid gardener. Comments furthered the growth of professional classes to advise on and design for one or the other type of building or landscape or its user. As a result, the organic quality of the environment, whether natural or built, was established by the implication of complex series of causal relations between figures and the room or garden through which they moved, within which they observed living and non-living things and where they acquired powers of discernment and a sense of taste – an ensemble in which the language of functions and purposes played a coordinating, unifying role. We still speak this language today.

Consigned to a nature seen as distinct from the workings of humans, plants remain mute to the techniques that manipulate their environment in the glasshouse, terrarium or biome just as mice still tunnel silently through loaves of bread, presumably unaware and certainly unappreciated, unlike Sir Isambard Brunel in his day. Distinctions between ourselves and other creatures remain as important now as they were for the vegetalist and the alchemist. They mean different things today; hence contemporary anxiety concerning genetic engineering that puts carrot genes in salmon to make it appear redder, vitamin A in 'golden rice' to make children in Asia see better, and efforts to create in our fields, houses and gardens zones entirely free of genetically modified organisms or to preserve indigenous plant species before they are lost forever. However, the object of technological interventions – life – escapes total manipulation. Unitary theories of the cell and such notions as environmental determinism have failed to capture the self-reflexive consciousness of the being that is the source of all forms of perception and self-reflection, although evolutionary psychologists may try.

Concerning perceptions of the order versus the appearance of things, which so intrigued figures like John Claudius Loudon, Gottfried Semper, James Fergusson and William Robinson in the nineteenth century and then Nikolaus Pesvner in the twentieth, one can conclude that, on the one hand, there exists the idea of 'function' as it comes to us from the sciences, particularly biological science – a world of means and ends, causes and effects that surrounds us all. When applied to various domains of human activity it is one term in a highly useful vocabulary for describing things and discriminating between them. On the other hand, one can think of 'functionalism' as forming a theoretical trend, a school of thought or a style of reasoning practised or 'owned' by a given profession or design practitioner. There is a category of means by which needs are recognized as such by the human sciences – a source of problems to which architects and landscape architects respond in various ways – and then there is a manner of thinking about the nature of these means, of deriving sense from them. Encompassing a broad range of technologies for representing basic biological necessities for shelter, light or warmth and for making sense of our own position in the order of things, the built environment is perhaps less distinct to the natural environment as a separate domain of existence, but is rather a means of making nature 'real'.

Looking back on the accounts of life and death in the glasshouse and environmental awareness in the Victorian home, a sense of experimentation and inventiveness as well as insecurity pervades the circumstances detailed in previous chapters. It is a sense that invokes the Janus-like identity of humankind, the two roles played by what Foucault termed 'man', or two of what Kant called 'certain great dispositions in our nature'. On the one hand, the belief in the unique self-consciousness of the human species makes it the source or systematizer of sensibility, experience and perception. Though sharing our artificial climates with 'harmless animals' as well as plants, it is the human species alone that understands – or believes that it does – why they are all there together in the first place or why it is too dark, too hot or cold, smoky within or foggy outside. It is the human being that is able to open or close a window if need be or devise some other means of improving its living arrangements. On the other hand, this being, like any other living creature, is subject to the functions of spaces, causes and effects that this understanding reveals. Humans too benefit from an 'atmosphere balmy and delicious' or suffer from tropical oppressiveness, though in different ways to other species. Unlike the reasoning subject of the camera obscura, our understanding of psychology today suggests that the mind is all too easily twisted towards decrepitude like the quaintly twisted columns of the conservatory – the conservatory, a mimicry of nature; the camera obscura an aberration of a different kind. Our dissatisfaction with our homes today is part of 'our' nature. Having construed the world of living and non-living things in such complex terms and having devised a way of measuring our intellectual abilities, it is easier to assert our ignorance of the ecosytems of which houses and gardens are an inextricable part. It is commonly said that 'the more we know the less we know'. With the emergence of biology and related disciplines the object of understanding, namely life – reducible to no single attribute or state of being – has become not only the source of stimuli, but has also come to afford material for perception and provides the basis of various norms. I have called this object the environment. It suffices to say that with doubled faces, looking back to our kinship with a universe of living creatures and forward to face the consequences, it is possible to inhabit the world – or at least, to consider our options for doing so.

Notes

First publication date is indicated by a subscript '$_1$'.

Preface

1. Alan Bullock and Oliver Stallybrass (eds), *The Fontana Dictionary of Modern Thought*, London: Fontana, 1983, $_1$1977, p. 207.

2. George Hersey, 'J.C. Loudon and Architectural Associationism', in *Architectural Review*, 144, August 1968, p. 89.

3. Melanie Louise Simo, *Loudon and the Landscape*, New Haven, CT: Yale University Press, 1988, pp. 247–9.

4. George Mackenzie, 'On the Form Which the Glass of a Forcing-House Ought to Have, in Order to Receive the Greatest Possible Quantity of Raysfrom the Sun' (1815), in *Transactions of the Horticultural Society of London*, II, London, 1817, pp. 171–5.

5. A useful history of the Palm House, its collections and reconstructions, is found in Sue Minter's *The Greatest Glass House*, London: HMSO, 1990.

6. Nikolaus Pevsner, *Pioneers of Modern Design*, Harmondsworth: Penguin, 1975, $_1$1936, p. 132.

7. Asa Briggs, *Iron Bridge to Crystal Palace: Impact and Images of the Industrial Revolution*, London: Thames and Hudson, 1979.

Introduction

1. Cited by John Hix in *The Glasshouse*, London: Phaidon, 1996, $_1$1974, pp. 13–14.

2. Joseph Dalton Hooker, letter to the First Commissioner of Works, 8 August 1867; excerpt in John Smith, *History of the Royal Gardens at Kew*, Royal Botanic Gardens, Kew: *Kewensia* Collection, 1880, p. 247.

3. Alchemy was a subject of curiosity and prompted the regular exchange of ideas between two of the great names of early modern science, Robert Boyle and Isaac Newton, as recently as the seventeenth century; see Lisa Jardine, *Ingenious Pursuits*, London: Little, Brown and Co., 1999, pp. 326–31.

4. Mary Woods and Arete Warren blame the tinted glass originally used in the Palm House, a conclusion formed earlier by John Hix. Minter offers substantial proof that the heating system was inadequate. See Woods and Warren, *Glasshouses*, New York: Rizzoli, 1988, p. 127; Hix, *The Glasshouse*, p. 148, and Sue Minter, *The Greatest Glass House*, London: HMSO, 1990, pp. 6–7.

5. In an early and celebrated work, *The Order of Things*, Michel Foucault described a significant 'mutation' in thought at the end of the eighteenth century corresponding to the appearance of the sciences of biology, economics and linguistics, famously announcing the arrival of 'Man' – the subject of human science – and a 'recent invention'. Following this transformation, anthropology, sociology and psychology in their contemporary forms sought to describe what it meant to be a human being. See Foucault, *The Order of Things: an Archaeology of the Human Sciences*, New York: Vintage Books Edition, 1973, $_1$1966, pp. 344–87.

6. Hix, *The Glasshouse*, pp. 13–14.

7. Woods and Warren, *Glasshouses*, p. 1.

8. A work that emphasizes less the origins of any one building type per se, but rather the role of scientific innovation in the history of the built environment is Reyner Banham's *The Architecture of the Well-Tempered Environment*, London: Architectural Press, 1969. Describing developments in artificial lighting, heating, ventilation and air-conditioning, and other building services, Banham does not explore the notion of environmental awareness in terms used here.

9. John Claudius Loudon, *An Encyclopaedia of Gardening*, London, 1822, p. 1012.

10. John Archer, *The Literature of British Domestic Architecture, 1715–1842*, Cambridge, MA: The MIT Press, 1985, pp. 46–56.

11. Ray Desmond, 'Who Designed the Palm House in Kew Gardens?', in *Kew Bulletin*, 27:2, 1972; Edward Diestelkamp, 'The Design and Building of the Palm House, Royal Botanic Gardens, Kew', in *Journal of Garden History*, 2:3, July–September 1982. John Hix cites these and other attempts to attribute responsibility for the design of the building. A similar controversy surrounding the design of the Winter Garden at Regent's Park (1846) involved the same cast of characters, Burton and Turner; see Hix, *The Glasshouse*, pp. 142–3, note 6.

12. Of course, the origins of functionalist thinking have long interested theorists and historians. The literature on functionalism in architecture is extensive. For different takes on the subject, see Edward DeZurko, *Origins of Functionalist Theory*, New York: Columbia University Press, 1957; Alan Colquhoun, *Modernity and the Classical Tradition: Architectural Essays 1980–1987*, Cambridge, MA: The MIT Press, 1989; and Adolf Behne, *The Modern Functional Building*, Santa Monica, CA: Getty Research Institute for the History of Art and the Humanities, 1996.

13. Efforts to derive sense from spaces, functional or otherwise, raise a particularly difficult issue regarding ideas and their representation. Ian Hacking has argued that the 'problem' of reason, ideas and their representation is relatively of little consequence, being mainly a concern by the historian of science for philosophical issues like the nature of mind and reality, objectivity and truth. Countering a 'philosophy' of science as an investigation of the workings of reason in the world, his work has entailed a 'history' of science as an ensemble of practices involving those of 'intervention and action and experiment' rather than 'representation and thinking and theory'. See Ian Hacking, *Representing and Intervening*, Cambridge: Cambridge University Press, 1983.

14. Thomas Knight, *The Report of a Committee of the Horticultural Society of London*, London, 1805, pp. 4–5.

15. Michel Foucault, *Madness and Civilization: a History of Insanity in the Age of Reason*, 1973, $_1$1965; *The Birth of the Clinic: an Archaeology of Medical Perception*, 1973, $_1$1963; *Discipline and Punish: the Birth of the Prison*, 1977, $_1$1975.

16. Canguilhem described the life sciences as developing as both an ensemble of knowledges and of practices – technologies for managing tensions between humans and their surroundings. By relating the health of human beings to the state of their environs, biology and related disciplines opened a domain of moral and ethical concerns for the 'problem' of life and death. Delaporte demonstrated that a consistent feature of scientific practice over the past several centuries has been to circumscribe the study of plants with that of animals and the study of animals with human beings, so that observations of one kingdom offered lessons on how the other worked. Knowledge of vegetable life prefigured an understanding of animal physiology and provoked broader concerns forming the life sciences, while the human sciences arose with the task of articulating the meaning these various studies held for humankind. Where the former ask 'how' we live, the latter asks 'why'. It follows that the history of changing attitudes to nature is likewise a study of changing ideas regarding the uniqueness of species and humans in particular. See Georges Canguilhem, *The Normal and the Pathological*, New York: Zone Books, 1989, $_1$1943, p. 20; and François Delaporte, *Nature's Second Kingdom: Explorations of Vegetality in the Eighteenth Century*, Cambridge, MA: The MIT Press, 1982.

17. Jonathan Crary, *Techniques of the Observer: on Vision and Modernity in the Nineteenth Century*, Cambridge, MA: The MIT Press, 1990.

18. Richard Grove, *Green Imperialism: Colonial Expansion, Tropical Island Edens, and the Origins of Environmentalism, 1600–1860*, Cambridge: Cambridge University Press, 1995, p. 14.

19. Another useful biography is John Gloag's *Mr. Loudon's England: the Life and Work of John Claudius Loudon, and his Influence on Architecture and Furniture Design*, Newcastle upon Tyne: Oriel Press, 1970. A collection of essays that addresses the context for many of Loudon's ideas is Elisabeth MacDougall (ed.), *John Claudius Loudon and the Early Nineteenth Century in Great Britain*, Washington, DC: Dumbarton Oaks Trustees for Harvard University, 1980.

20. Describing this socio-psychological process as the 'secularisation of the soul', Ian Hacking wrote that, from a modern perspective, 'To think of the soul is not to imply that there is one essence, one spiritual point, from which all voices issue. In my way of thinking the soul is a more modest concept than that. It stands for the strange mix of aspects of a person that may be, at some time, imagined as inner.' See Hacking, *Rewriting the Soul, Multiple Personality and the Sciences of Memory*, Princeton, NJ: Princeton University Press, 1995, pp. 5–6.

Chapter 1

1. Arthur Lovejoy, *The Great Chain of Being: a Study of the History of an Idea*, Cambridge, MA: Harvard University Press, 1936, pp. 227–41.

2. Intentions behind these evocations of Eden and paradise gardens are varied, though a sample of works includes: John Prest, *The Garden of Eden: the Botanic Garden and the Re-Creation of Paradise*, New Haven, CT: Yale University Press, 1981; David Bourdon, *Designing the Earth: the Human Impulse to Shape Nature*, New York: Abrams, 1995; and Joseph Rykwert, *On Adam's House in Paradise*, New York: The Museum of Modern Art, 1972.

3. Edward Dudley and Maximillian Novak (eds), *The Wild Man Within: an Image in Western Thought from the Renaissance to Romanticism*, Pittsburgh, PA: University of Pittsburgh Press, 1972; Roger Bartra, *Wild Men in the Looking Glass: the Mythic Origins of European Otherness*, Ann Arbor, MI: University of Michigan Press, 1994.

4. Paul Hirst and Penny Woolley, 'Nature and Culture in Social Science: the Demarcation of Domains of Being in Eighteenth Century and Modern Discourses', in *Geoforum*, 16:2, 1985; also of interest by Hirst and Woolley is *Social Relations and Human Attributes*, London: Tavistock, 1982.

5. Jan Nieuhof, *An Embassy from the East India Company of the United Provinces to the Grand Tartar Cham, Emperor of China*, trans. from the Latin edn by John Ogilby, London, 1669, p. 165. This passage and other observations are cited by Patrick Conner in *Oriental Architecture in the West*, London: Thames and Hudson, 1979.

6. Louis Le Comte, cited in Conner, *Oriental Architecture*, p. 21.

7. Lord George Macartney, *An Embassy to China: Being the Journal Kept by Lord Macartney During his Embassy to the Emperor Ch'ien-lung 1793–1794* (1797), edited with an Introduction and Notes by Launcelot Cranmer-Byng, London: Longman, 1962, p. 224.

8. Ibid., pp. 224–67.

9. Ibid., p. 271.

10. Ibid., p. 223.

11. Cited in Thomas Hughes (ed.), *Selections from 'Lives of the Engineers'*, Cambridge, MA: The MIT Press, 1966, p. 11.

12. Foucault's analysis is admittedly problematic for he attempts to locate meaning at a fundamental level of representation, assuming this to be an essential aspect of the human condition. The concept of the episteme in Foucault's intellectual history is addressed in Mark Cousins and Athar Hussain, *Michel Foucault: Theoretical Traditions in the Social Sciences*, London: Macmillan, 1984. For a review of the structuralist underpinnings of the concept and discussion of Foucault's archaeological investigations versus his subsequent genealogical methods, see Phil Bevis, Michele Cohen and Gavin Kendall, 'Archaeologizing Genealogy: Michel Foucault and the Economy of Austerity', in Mike Gane and Terry Johnson (eds), *Foucault's New Domains*, London: Routledge, 1993, and Beverley Brown and Mark Cousins, 'The Linguistic Fault: the Case of Foucault's Archaeology', in *Economy and Society*, 9:3, August 1980.

13. Michel Foucault, *The Order of Things: an Archaeology of the Human Sciences*, New York: Vintage Books Edition, 1973, ₁1966, p. 17.

14. Chronicles, such as William Harrison's *Description of England* (first printed in Raphael Holinshed's *Chronicles* of 1577) can be compared to the genre of *histoires* including Aldrovandi's *Monstrorum Historia* (1647) in which observation is often freely mixed with anecdote and myth in rambling fashion. Other, practical manuals are worth considering. Thomas Tusser's *One Hundred Points of Good Husbandry* (1557) is typical and sets out in one hundred passages of four-line verse instructions to the farmer for establishing order in the garden and household, an order prerequisite to a virtuous life. Ellen Eyler's *Early English Gardens and Garden Books*, Ithaca, NY: Cornell University Press, 1963, provides a useful account of such early books.

15. Foucault, *The Order of Things*, p. 129.

16. Mircea Eliade, *The Forge and the Crucible*, London: Rider, 1962, pp. 11, 169.

17. James Lackington, *Lives of the Alchemists*, London, 1815, pp. 13–14.

18. Peter Bowler, *The Fontana History of the Environmental Sciences*, London: Fontana Press, 1992, p. 61.

19. Antonio Filarete, cited in Rudolf Wittkower, *Architectural Principles in the Age of Humanism*, 4th edn, London: Academy Editions, 1973, ₁1949, p. 10.

20. Paul Hirst, 'Foucault and Architecture', in *AA Files*, 26, Autumn 1993.

21. Foucault, *The Order of Things*, p. 51.

22. Lovejoy, *The Great Chain of Being*, pp. 227–41.

23. Nicholas Jardine, James Secord and Emma Spary (eds), *Cultures of Natural History*, Cambridge:

Cambridge University Press, 1996.

24. François Delaporte, *Nature's Second Kingdom: Explorations of Vegetality in the Eighteenth Century*, Cambridge, MA: The MIT Press, 1982, p. 187.

25. The study was not entirely new to the seventeenth and eighteenth centuries, as Aristotle formerly had proposed that life was dependent upon the vegetative power (*potentia vegetative*) that enabled creatures to thrive and multiply. This power was distinguished from the capacities to sense, to desire and obviously to think. Subsequent studies of the manner in which living beings functioned and thrived were commonly expositions of the actions of the soul on the body, just as systems of classification arose, in part, to elaborate the account of creation found in Genesis. See Nicholas Stenick, *Science and Creation in the Middle Ages*, Notre Dame, IN: University of Notre Dame Press, 1976, pp. 105–20. The study of 'animality' was the logical counterpart to vegetable science – denoting the 'whatever it is' unique to animals. For the sake of clarity, the term will not be used here.

26. David Hume, *Dialogues Concerning Natural Religion*, London, 1779, $_1$1751, Part VII.

27. François Jacob, *The Logic of Living Systems: a History of Heredity*, London: Allen Lane, 1974, $_1$1970, cited in Delaporte, *Nature's Second Kingdom*, p. 13.

28. Georges Canguilhem, *Ideology and Rationality in the History of the Life Sciences*, Cambridge, MA: The MIT Press, 1988, pp. 84–5.

29. Cited in Robert Harvey-Gibson, *Outlines of the History of Botany*, New York: Arno Press, 1981, $_1$1919, p. 72.

30. From *Eloges des Académiciens* (Paris, 1799), cited in Delaporte, *Nature's Second Kingdom*, p. 136.

31. Ibid., p. 185.

32. Hirst and Woolley, 'Nature and Culture', p. 152.

33. Jonathan Swift, *Travels into Several Remote Nations of the World*, London, 1726.

34. Jean-Jacques Rousseau, *Discourse on the Origin of Inequality Among Men* ($_1$1755), in George Cole (trans.), *The Social Contract, and Discourses*, London: Dent, 1986.

35. Hirst and Woolley, 'Nature and Culture', p. 153.

36. Jean Starobinski (1962, $_1$1943) cited in Georges Canguilhem, *The Normal and the Pathological*, New York: Zone Books, 1989, p. 242.

37. Daniel Defoe (1787), cited in Dudley and Novak, *The Wild Man Within*, p. 196.

38. Edgar Rice Burroughs, *Tarzan of the Apes*, New York, 1912.

39. Marc-Antoine Laugier, *An Essay on Architecture*, trans. with Introduction by Wolfgang and Anni Hermann, Los Angeles: Hennessey and Ingalls, 1977, $_1$1753, p. 11–12.

40. Ibid., p. xx.

41. Jean-Jacques Rousseau, *The Social Contract* ($_1$1762), cited in Cole, *The Social Contract, and Discourses*, p. 169.

42. For criticisms of Laugier's position by his contemporaries, see Anthony Vidler, 'Rebuilding the Primitive Hut', *The Writing of the Walls: Architectural Theory in the Late Enlightenment*, Princeton, NJ: Princeton Architectural Press, 1987, p. 21. Le Corbusier built upon the hypothetical character of the primitive hut and the character of the human builder in *Towards a New Architecture*, trans. from French by Frederick Etchells, London: Architectural Press, 1946, $_1$1931, pp. 68–9.

43. Sylvia Lavin, 'The History of Ethnography', *Quatremère de Quincy and the Invention of a Modern Language of Architecture*, Cambridge, MA: The MIT Press, 1992, pp. 62–75.

44. Cited in Cole, *The Social Contract, and Discourses*, p. 85.

45. For some the divide between the utility and appearance of form suggests a crisis in modern thought itself, demanding either the reconciliation of these terms through recourse to phenomenology or other philosophical means. See Alberto Perez-Gomez, *Architecture and the Crises of Modern Science*, Cambridge, MA: The MIT Press, 1983 and relatedly, Dalibor Vesely, 'Architecture and the Conflict of Representation', in *AA Files*, 8, January 1985.

46. Nikolaus Pevsner, *Pioneers of Modern Design*, Harmondsworth: Penguin, 1975, $_1$1936, p. 128.

47. Pevsner's comments are likewise a valuation of the illusory nature of professional titles. Brunel's status as an engineer, unlike his susceptibility to beauty, was simply accidental, secondary to his own integrity as a discerning designer.

48. Gottfried Semper, 'The Four Elements of Architecture', *The Four Elements of Architecture and Other Writings* (1815), trans. with an Introduction by Harry Mallgrave and Wolfgang Hermann, Cambridge: Cambridge University Press, 1989.

49. Harry Mallgrave, *Gottfried Semper: Architect of the Nineteenth Century*, New Haven, CT: Yale University Press, 1996, pp. 179–89.

50. James Fergusson, *A History of Architecture in All Countries from the Earliest Times to the Present Day*, 3rd edn, London: John Murray, 1893, p. 13.

51. John Dixon Hunt, *The Figure in the Landscape: Poetry, Painting, and Gardening During the Eighteenth Century*, Baltimore, MD: Johns Hopkins University Press, 1976, p. 63.

52. John Claudius Loudon, *An Encyclopaedia of Gardening*, London, 1822, p. 111.

53. William Robinson, *Hardy Flowers*, London, 1871, p. xxxiii.

54. Robin Fox in an Introduction to the reprint of the 4th edn (1894) of William Robinson, *The Wild Garden* (London: Scolar Press, 1977, ₁1870), pp. xxvi, xxi; relevant handbooks included Richard Prior, *On the Popular Names of British Plants, Being an Explanation of the Origin and Meaning of the Names of Our Indigenous and Most Commonly Cultivated Species*, London, 1870; James Britten and Robert Holland, *A Dictionary of English Plant-Names*, London, 1886.

55. Alan Tate, 'Loudon and the Return to Formality', in Elisabeth MacDougall (ed.), *John Claudius Loudon and the Early Nineteenth Century in Great Britain*, Washington, DC: Dumbarton Oaks Trustees for Harvard University, 1980, pp. 59–76.

56. John Claudius Loudon, *Sketches of Hot-Houses* (1818), a manuscript cited in Melanie Louise Simo, *Loudon and the Landscape*, New Haven, CT: Yale University Press, 1988, p. 112; Loudon's designs were detailed in *Remarks on the Construction of Hothouses*, London, 1817.

57. Loudon, *An Encyclopaedia of Gardening*, p. 1013.

58. George Hersey, 'J.C. Loudon and Architectural Associationism', in *Architectural Review*, 144, August 1968, p. 89; and *High Victorian Gothic: a Study in Associationism*, Baltimore, MD: Johns Hopkins University Press, 1972.

59. Robinson, *The Wild Garden*, p. 93

60. Reyner Banham's *The Architecture of the Well-Tempered Environment*, London: Architectural Press, 1969.

61. George Romanes, 'Animal Intelligence', in *Fortnightly Review*, December 1881, p. 740 (source: *Oxford English Dictionary*).

Chapter 2

1. John Hammond, *The Camera Obscura*, Bristol: Hilger, 1981.

2. Jonathan Crary, *Techniques of the Observer: on Vision and Modernity in the Nineteenth Century*, Cambridge, MA: The MIT Press, 1990; Hal Foster (ed.), *Vision and Visuality*, Seattle, WA: Bay Press, 1988.

3. François Delaporte, *Nature's Second Kingdom: Explorations of Vegetality in the Eighteenth Century*, Cambridge, MA: The MIT Press, 1982, p. ix.

4. Stephen Hales, *Vegetable Staticks, or an Account of Some Statical Experiments on the Sap in Vegetables*, London, 1727, p. 1.

5. Marcello Malpighi (1686), cited in Delaporte, *Nature's Second Kingdom*, pp. 39–40.

6. Hales, *Vegetable Staticks*, pp. 2–3.

7. Ibid., p. 136.

8. Jean-Etienne Guettard (1748), cited in Delaporte, *Nature's Second Kingdom*, p. 61.

9. John Hill, 'The Sleep of Plants' (1757), *Botanical Tracts*, London, 1762, section x.

10. Hill, *Botanical Tracts*, p. 29.

11. Duhamel du Monceau (1751), cited in Delaporte, *Nature's Second Kingdom*, p. 169.

12. Thomas Knight, *The Report of a Committee of the Horticultural Society of London*, London, 1805.

13. Delaporte, *Nature's Second Kingdom*, p. 61.

14. Ibid., p. 60.

15. George Tod, *Plans, Elevations and Sections of Hothouses, Green-Houses, an Aquarium, Conservatories, etc.*, London, 1812, p. 5.

16. George Mackenzie, 'On the Form which the Glass of a Forcing-house Ought to Have, in Order to Receive the Greatest Possible Quantity of Rays From the Sun' (1815), in *Transactions of the Horticultural Society of London*, II, London, 1817, pp. 171–5.

17. John Claudius Loudon, *Remarks on the Construction of Hothouses*, London, 1817, p. 4.

18. John Claudius Loudon, *An Encyclopaedia of Gardening*, London, 1822, p. 584.

19. Mackenzie, 'On the Form Which the Glass of a Forcing-house Ought to Have', p. 174. See also George Mackenzie, 'Description of an Economical Hot-House' (1814), *Memoires of the Caledonian Horticultural Society*, II, Edinburgh, 1818, pp. 55–6.

20. John Claudius Loudon, *A Short Treatise on Several Improvements, Recently Made in Hot-Houses*, Edinburgh, 1805, pp. 39–43.

21. John Locke, *An Essay Concerning Human Understanding*, II, London, 1748, ₁1690, xi; cited in Crary, *Techniques of the Observer*, pp. 41–2.

22. William Wollaston, 'On a Periscopic Camera Obscura and the Microscope', in *Philosophical Transactions of the Royal Society of London*, 102, London, 1812, pp. 370–77. Barbara Stafford discusses the problem of ambiguity of microscopic imagery for eighteenth-century natural historians. The issue encouraged, amongst other responses, the further specialization of researchers and the development of a more abstract language to describe what was being seen beneath the lens. See Stafford, 'Images of Ambiguity', in David Miller and Peter Reill (eds), *Visions of Empire: Voyages, Botany and Representations of Nature*, Cambridge: Cambridge University Press, 1996, pp. 230–57.

23. Humphry Davy, *Elements of Agricultural Chemistry*, London, 1813, p. 34.

24. Robert Harvey-Gibson, *Outlines of the History of Botany*, New York: Arno Press, 1981, ₁1919, pp. 83–7.

25. Locke, *An Essay Concerning Human Understanding*, I, i. §1. Suggesting the similarity of reason and experience on the basis of a causal force that supported both, David Hume nonetheless found it harder to discern evidence of reason and design in the natural world than mere generative or vegetable forces, for the causes of the former were more complex than the latter. For Hume, like Locke, reason was distinct to the machinations of experience. See David Hume, *Dialogues Concerning Natural Religion*, VII, London, 1779, ₁1751.

26. Alain Corbin, *The Foul and the Fragrant: Odor and the French Social Imagination*, Leamington Spa: Berg, 1986, p. 106.

27. Stephen Hales, *A Description of Ventilators: Whereby Great Quantities of Fresh Air May with Ease be Conveyed into Mines, Gaols, Hospitals, Work-Houses and Ships, in Exchange for their Noxious Air, etc.* London, 1743, pp. iv–v.

28. Georgius Agricola, *De Re Metallica*, Basel, 1556.

29. Charles Thackrah, *The Effects of the Principal Arts, Trades, and Professions, and of Civic States and Habits of Living, on Health and Longevity, etc.*, London, 1831.

30. Cuthbert Johnson, *On the Cottages of Agricultural Labourers: with Economical Working Plans, and Estimates for their Improved Construction*, London, 1847. There are many titles comprising this genre. Some were directed to the landed gentry keen to improve the ornamental value of their estates with refurbished cottages and, if convenient and economic, the lives of their workers; see John Papworth, *Rural Residences: Consisting of a Series of Designs for Cottages, Decorated Cottages, Small Villas, and Other Ornamental Buildings*, London, 1818.

31. Robin Evans, *The Fabrication of Virtue: English Prison Architecture, 1750–1840*, Cambridge: Cambridge University Press, 1982, p. 95.

32. As Delaporte observed: 'In its place a scientific practice was elaborated in accordance with different rules and related in a new way to animal physiology. The plant in effect became the laboratory of cellular theory. The tables were turned; from this point on, the lower forms of life were to shed light on the higher. Of course the object of study was now quite different, interest centering not on organs and functions but on the intimate structure of the plant, its fundamental composition.' in *Nature's Second Kingdom*, p. 189.

33. Alan Morton, *History of Botanical Science: an Account of theDevelopment of Botany from Ancient Times to the Present Day*, London: Academic Press, pp. 287–8.

34. Harvey-Gibson, *Outlines of the History of Botany*, p. 103.

35. Morton, *History*, pp. 328–9.

36. Ibid., pp. 336–7.

37. Cited by Stephen Ward, *Wardian Cases for Plants and their Application*, London, 1854, p. 22.

38. Ian Hacking, *Representing and Intervening*, Cambridge: Cambridge University Press, 1983, pp. 156–7.

39. Davy, *Elements of Agricultural Chemistry*, p. 9.

40. In doing so he rejected Xavier Bichat's theory of vitalism which had implied that physiology was not reducible to the laws of chemistry and physics. He argued, rather, that it was reducible and that physiological phenomena were predictable and experimental methods allowed them to be understood. See Robert Young, *Mind, Brain and Adaptation in the Nineteenth Century*, Oxford: Clarendon Press, 1970, p. 73.

41. Ibid., p. 93.

42. Ibid., p. 121.

43. Thomas Knight, 'On the Ill Effects of Excessive Heat in Forcing-houses During the Night' (1814), in *Transactions of the Horticultural Society of London*, II, London, 1817, p. 131.

44. Hill, 'The Sleep of Plants'.

45. Knight, 'On the Ill Effects of Excessive Heat', p. 130.

46. Bruno Latour, *Science in Action*, Cambridge, MA: Harvard University Press, 1987, p. 229.

47. Michel Foucault, *The Order of Things: an Archaeology of the Human Sciences*, New York: Vintage, 1973, ₁1966, p. 232.

48. Harvey-Gibson, *Outlines of the History of Botany*, p. 62.

49. Jan Ingenhousz, *Experiments upon Vegetables, Discovering their Great Power of Purifying the Common Air in Sunshine, and of Injuring it in the Shade and at Night*, London, 1779.

50. For Harvey-Gibson: 'The discovery of Dr. Priestley that plants thrive better in foul air than in dephlogistonated air, and that plants have a power of correcting foul air, has thrown a new and important light upon the arrangements of this world. It shows, even to a demonstration, that the vegetable kingdom is subservient to the animal: and, vice-versa, that the air, spoiled and rendered noxious to animals by their breathing in it, serves to plants as a kind of nourishment', *Outlines*, p. 64.

51. With its appearance as an object of study, biological or life science reversed the subservient relationship which had existed between Classical vegetality and taxonomy. In the earlier era, the observation of the formal structure and function of living beings was used to create a science of physiology. Now, the organic structure of beings – a totality which 'weaves back into the unique fabric of its sovereignty both the visible and the invisible' – became the measure of life. Foucault, *The Order of Things*, p. 229.

52. Knight, *The Report of a Committee*, p. 1.

Chapter 3

1. Melanie Louise Simo, *Loudon and the Landscape*, New Haven, CT: Yale University Press, 1988, pp. 247–49.

2. John Claudius Loudon, *Remarks on the Construction of Hothouses*, London, 1817, p. 4.

3. Philip Gosse, *Wanderings Through the Conservatories at Kew*, London, 1857, p. 44.

4. Henry Home, *Gentleman Farmer*, Edinburgh, 1776, p. viii; John Williams, 'On the Cultivation of the Vine in Forcing-Houses, with Observations on Forcing Peaches', in *Transactions of the Horticultural Society of London*, II, London, 1814, p. 108.

5. George Warner, *Landmarks in English Industrial History*, London: Blackie and Son, 1924, ₁1899, p. 219.

6. Werner Stark, *Jeremy Bentham's Economic Writings*, I, London: Allen and Unwin, 1952, p. 206.

7. Colin Russell, *Science and Social Change*, London: Macmillan, 1983, p. 91.

8. Ibid., p. 94; Simo notes the particular influence on Loudon of Professor Andrew Coventry in *Loudon and the Landscape*, p. 4.

9. This period is well documented by John Gloag in *Mr. Loudon's England: the Life and Work of John Claudius Loudon, and his Influence on Architecture and Furniture Design*, Newcastle upon Tyne: Oriel Press, 1970.

10. John Loudon, *A Short Treatise on Several Improvements, Recently Made in Hot-Houses*, Edinburgh, 1805.

11. This period of experimentation is detailed in Simo, *Loudon and the Landscape*, pp. 111–18; and George Kohlmaier, *Houses of Glass: a Nineteenth Century Building Type*, Cambridge, MA: The MIT Press, 1986.

12. George Mackenzie, 'On the Form which the Glass of a Forcing-House Ought to Have, in Order to Receive the Greatest Possible Quantity of Rays from the Sun' (1815), in *Transactions of the Horticultural Society of London*, II, London, 1817, p. 172.

13. Carlo Cipolla (ed.), *The Emergence of Industrial Societies*, London: Fontana, 1973, p. 181.

14. Stark, *Jeremy Bentham's Economic Writings*, p. 262.

15. Harold Dutton, *The Patent System and Inventive Activity During the Industrial Revolution, 1750–1852*, Manchester: Manchester University Press, 1984, p. 65.

16. Gloag, *Mr. Loudon's England*, p. 5.

17. John Claudius Loudon, *An Encyclopaedia of Gardening*, 5th edn, London, 1835, pp. 1012–13.

18. Samuel Coleridge, *General Introduction to the Encyclopaedia Metropolitana; or, A Preliminary Treatise on Method*, London, 1818.

19. Ibid., p. 15.

20. Loudon, *An Encyclopaedia of Gardening*, p. 1164.

21. Stephen Hales, *Vegetable Staticks, or an Account of Some Statical Experiments on the Sap in Vegetables*, London, 1727.

22. John James Stevenson, *House Architecture*, London, 1880, p. 280.

23. Barbara Stafford suggests this was one consequence of the growth of microscopy; see Stafford, 'Images of Ambiguity', in David Miller and Peter Reill (eds), *Visions of Empire: Voyages, Botany and Representations of Nature*, Cambridge: Cambridge University Press, 1996, p. 254.

24. Immanuel Kant wrote: 'In metaphysical speculations, it has always been assumed that all our knowledge must conform to objects; but every attempt from this point of view to extend our knowledge of objects *a priori* by means of conceptions had ended in failure. The time has now come to ask, whether better progress may not be made by supposing that objects must conform to our knowledge.' See Kant, *Critique of Pure Reason*, trans. by J. Bernard, New York: Hafner Publishing Company, 1965, $_1$1781, p. xvi.

25. Ian Hacking addressed the existence of 'styles' of reason, taking the word from *Styles of Scientific Thinking in the European Tradition* (1983), where Alistair Combie described six distinct forms of investigation which have characterized the development of scientific knowledge. Hacking described the 'hypothetico-deductive' as arising in part as a response to the anti-realist reactions to Kant's noumenal world, where there exist 'things-in-themselves' that underlie experience of both physical nature and our own mental states. See Hacking, *Representing and Intervening*, Cambridge: Cambridge University Press, 1983, pp. 1, 99.

26. Donal McCracken, *Gardens of Empire: Botanical Institutions of the Victorian British Empire*, London: Leicester University Press, 1997; Madeleine Bingham, *The Making of Kew*, London: Michael Joseph/Folio Society, 1975; Mea Allan, *The Hookers of Kew, 1785–1911*, London: Michael Joseph/Folio Society, 1967.

27. Edward Diestelkamp, 'The Design and Building of the Palm House, Royal Botanic Gardens, Kew', in *Journal of Garden History*, 2:3, July–September 1982. The excerpts of correspondence that follow are more fully elaborated and their sources precisely cited in this article.

28. Ibid., p. 235.

29. Ibid., p. 236.

30. George Chadwick questioned Burton's contribution at Chatsworth in *Works of Sir Joseph Paxton*, London: Architectural Press, 1961, p. 78.

31. Diestelkamp, 'The Design and Building', p. 237.

32. John Hix, *The Glasshouse*, London: Phaidon, 1996, $_1$1974, p. 139.

33. Diestelkamp, 'The Design and Building', p. 238.

34. Ibid.

35. Ibid.

36. Ibid.

37. Ibid.

38. Observation made by Thomas Drew in *Building News*, 19 March 1880, and noted by Diestelkamp, 'The Design and Building', p. 252.

39. Loudon, *An Encyclopaedia of Gardening*, p. 582.

40. Ibid., p. 587. Diestelkamp notes in 'The Design and Building' that the crown glass used almost exclusively during the eighteenth and nineteenth centuries had a slightly green tint caused by traces of copper. Clear glass, requiring greater effort to remove such impurities, was therefore more expensive.

41. The Palm House at Kew would have benefited from the April 1845 repeal of the tax levied on crown and sheet glass, after which prices were halved. It is uncertain the degree to which building economy dictated the exact choice of material in the end. The specifications for 595 gr (21 oz) per 30 cm (12 in) patent sheet glass for the gallery and roof of the palm stove and for 19 oz for other glass surfaces acknowledges the superior strength of sheet glass over crown and also its availability for larger glazing surfaces. As Hix records, sheet glass was flatter, with fewer surface irregularities that some thought was the actual source of scorching; Hix, *The Glasshouse*, p. 148.

42. Mackenzie, 'On the Form Which the Glass of a Forcing House Ought to Have', p. 176.

43. Hix, *The Glasshouse*, p. 148.

44. Sue Minter, *The Greatest Glass House*, London: HMSO, 1990, pp. 6–7.

45. John Williams, 'On the Cultivation of the Vine in Forcing Houses, with Observations on Forcing Peaches', 1814, in *Transactions of the Horticultural Society of London*, II, London, 1817, p. 113.

46. Cited in Minter, *The Greatest Glass House*, p. 4.

47. Mary Woods and Arete Warren, *Glasshouses*, New York: Rizzoli, 1988, p. 127.

48. Cited by Will Howie, 'Restoration of the Temperate House', in *Architects' Journal*, 175:24, 16 June 1982, p. 49.

49. Alan Morton, *History of Botanical Science: an Account of the Development of Botany from Ancient Times to the Present Day*, London: Academic Press, 1981, pp. 362–4.

50. Ibid., pp. 377–8.

51. Hazel Conway, *People's Parks: the Development and Design of Victorian Parks*, Cambridge: Cambridge University Press, 1991.

52. Charles Smith, *Parks and Pleasure Grounds*, London: Reeve, 1852, p. 156.

53. John H. Brazell, *London Weather*, London: HMSO, 1968, pp. 1–2. All climatological data cited here are from this work.

54. Ibid., pp. 11–12.

55. Ibid., pp. 102, 103.

56. Anonymous correspondent, 'Green-Glass in Plant-Houses', in *Kew Bulletin*, February 1895, pp. 43–45. Minter observed that worsening air pollution as well as increased levels of shade caused by maturing specimens created very different conditions in the Palm House in the 1890s to those at the time it was completed; Minter, *The Greatest Glass House*, p. 20.

57. William Dallimore, *A Gardener's Reminiscences*, Royal Botanic Gardens, Kew: Kewensia Collection, unpublished manuscript, *c*. 1955, vol. 1, p. 276.

58. Joseph Dalton Hooker, *Report on the Progress and Condition of the Royal Gardens at Kew, During the Year 1866*, London, 1866, p. 1.

59. Robert Buchanan, *The Engineers: a History of the Engineering Profession in Britain, 1750–1914*, London: Jessica Kingsley, 1989, p. 14.

60. Henry Thomson, *The Choice of a Profession*, London, 1857.

Chapter 4

1. Peter Bowler, *The Fontana History of the Environmental Sciences*, London: Fontana Press, 1992, pp. 365–6.

2. This heightened sensitivity to the influence of environmental factors on organic but, more specifically, human life, has varied sources. For the historian of ideas, these may well be identified by a survey of attitudes to nature arising within various Western intellectual traditions. A useful reference is Max Oelschlaeger, *The Idea of Wilderness: from Prehistory to the Age of Ecology*, New Haven, CT: Yale University Press, 1991.

3. Lynda Nead, *Victorian Babylon: People, Streets, and Images in Nineteenth-Century London*, New Haven, CT: Yale University Press, 2000, pp. 27–8.

4. Paxton's was one of a number of radical schemes received by the Parliamentary Select Committee on Metropolitan Communications in 1854, including W. Mosley's 'Crystal Way', a glass-covered and shop-lined railway thoroughfare running from St Paul's Cathedral to Oxford Circus; Stephen Halliday, *The Great Stink of London*, Stroud, Gloucestershire: Sutton, 1999, p. 164.

5. Cuthbert Johnson, *Our House and Garden: What We See, and What We Do Not See In Them*, London, 1864, p. 144.

6. Thomas Cartwright, *Domestic Science. The Science of Domestic Economy and Hygiene Treated Experimentally*, London, 1900, p. 9.

7. Anon., *The House and its Surroundings*, London, 1878, p. 82.

8. A familiarity with these elements through philosophical or other, more 'practical' arts such as medicine or horticulture preceded their theoretical elaboration by nineteenth-century science. Subsequently, mechanical and chemical analyses were to recast relations between these elements and living matter. This occurred in a radically different way from that undertaken by the ancient Greeks or the common farmer, whose experience of such matters was more an outcome of experience and convention than scientific certainty.

9. John Claudius Loudon, *Suburban Gardener and Villa Companion*, London, 1838, pp. 177–80.

10. John Davy, *Lectures on the Study of Chemistry, in Connexion with the Atmosphere, the Earth, and the Ocean, etc.*, London, 1849.

11. Albert Howard, writing in the Preface to a re-issue of Charles Darwin's *The Formation of Vegetable Mould* (1881) entitled *Darwin on Humus and the Earthworm*, London: Faber and Faber, 1945.

12. John Fisher, *The Origins of Garden Plants*, London: Constable, 1982.

13. Cassell and Co., *Cassell's Household Guide: Being a Complete Encyclopoedia of Domestic and Social Economy: and Forming a Guide to Every Department of Practical Life*, I, London, 1869, p. 20.

14. Edward Girdlestone, *Our Debt and Duty to the Soil: or, The Poetry and Philosophy of Sewage Utilization*, London, 1878, p. 35.

15. Cassell and Co., *Cassell's Household Guide*, p. 21.

16. Christopher Davy, *The Architect, Engineer and Operative Builder's Constructive Manual, or, A Practical and Scientific Treatise on the Construction of Artificial Foundations for Buildings, Railways, &c.*, London, 1839.

17. James Cameron, *Is My House Healthy? How to Find Out*, Leeds, 1892, pp. 4–12.

18. Andrew Downing, *A Treatise on the Theory and Practice of Landscape Gardening Adapted to North America, With a View to the Improvement of Country Residences … With Remarks on Rural Architecture*, 6th edn, New York, 1859, ₁1841, p. 287.

19. George Wilkinson, *Practical Geology and Ancient Architecture*, Dublin, 1845, pp. 1, 2.

20. Ernest Cook, Introduction to John Ruskin, *The Stones of Venice*, London: George Allen, 1904, ₁1851, p. xlviii.

21. Robert Kerr, *The Gentleman's House*, 3rd edn, London, 1871, ₁1864, pp. 74, 75.

22. Robert Meikleham, *On the History and Art of Warming and Ventilating Rooms and Buildings, etc.*, London, 1845, II, p. 37.

23. David Ansted, *Natural History of the Inanimate Creation: Being a Guide to the Scenery of the Heavens, the Phenomena of the Atmosphere, the Structure and Geological Features of the Earth, and its Botanical Productions*, London, 1856, p. 10.

24. Vladimir Jankovic, *Reading the Skies: a Cultural History of English Weather, 1650–1820*, Chicago: The University of Chicago Press, 2000, pp. 165–7.

25. Eric Lampard, 'The Urbanizing World', in Harold Dyos and Michael Wolff (eds), *The Victorian City: Images and Realities*, London: Routledge, 1973. See also John Hassan, 'The Growth and Impact of the British Water Industry in the Nineteenth Century', in *The Economic History Review*, 38:4, November 1985, p 531.

26. Hassan, 'The Growth and Impact of the British Water Industry', pp. 532–3.

27. Cuthbert Johnson, *The Objects and History of the Thames Improvement Company*, London, 1839, p. 9.

28. John Davy, *Lectures*, p. 34.

29. Charles Singer (ed.), *A History of Technology*, Oxford: Oxford University Press, 1954–84, V, pp. 559–60; and Maurice Daumas (ed.), *A History of Technology and Invention*, III, New York: Crown, 1969, p. 457.

30. John Harris, 'Some Microscopical Observations of Vast Numbers of Animalcula Seen in Water', in *Philosophical Transactions*, London, 1695–97, 19, pp. 254–9.

31. Gustav Bischof, 'On Putrescent Organic Matter in Potable Water', in *Proceedings of the Royal Society of London*, 26, London, 1877, p. 152.

32. Cassell and Co., *Cassell's Household Guide*, p. 162,

33. Thoughts on the prevention of disease in the nineteenth century through means of sanitation were basically Hippocratic in origin, invoking an ancient ideal of pure soil, air and water. See James Cassedy, 'The Flamboyant Colonel Waring: An Anticontagionist Holds the American Stage in the Age of Pasteur and Koch', in Judith Leavitt and Ronald Numbers (eds), *Sickness and Health in America: Readings in the History of Medicine and Public Health*, Madison, WI: Wisconsin University Press, 1978, p. 307.

34. Barbara Abbott, 'Water = H₂O', *Mind*, 108:429, 1999, pp. 145–8.

35. A work that details other historical dimensions of this awareness is Alain Corbin's *Time, Desire, and Horror: Towards a History of the Senses*, Cambridge: Polity Press, 1995, ₁1991.

36. Cameron, *Is My House Healthy?*, p. 9.

37. Alain Corbin, *The Foul and the Fragrant: Odor and the French Social Imagination*, Leamington Spa: Berg, 1986, ₁1982.

38. Ann Sproule, *The Social Calendar*, Poole: Blandford Press, 1978, p. 29; and Morgan, Joan, and Richards, Alison, *A Paradise Out of a Common Field*, New York: Harper and Row, 1990, p. 23.

39. Esther Copley, *Catechism of Domestic Economy*, London, 1851, p. 22.

40. Testimony by a resident of Holborn, London, to a member of the Royal Commission on the Housing of the Working Classes in 1884, cited in Halliday, *The Great Stink*, p. 178.

41. Cited in Annemarie Adams, *Architecture in the Family Way*, Montreal: McGill–Queen's University Press, 1996, p. 36.

42. James Donald, *Imagining the Modern City*, London: Athlone Press, 1991, Chapter 1: 'Fog Everywhere'; Peter Ackroyd, *London, the Biography*, London: Chatto and Windus, 2000, Chapter 47: 'A Foggy Day'.

43. Cassell and Co., *Cassell's Household Guide*, p. 20.

44. Anon., 'Fogs in a Metropolitan Light', in *The Builder*, 23:1173, 29 July 1865, pp. 537–8.

45. Anon., *The Builder*, 52:2301, 12 March 1887, pp. 377–8.

46. Cassell and Co., *Cassell's Household Guide*, p. 312.

47. Ibid., p. 99.

48. Martin Danahay, 'The Aesthetics of Coal: Representing Soot, Dust, and Smoke in Nineteenth-Century Britain', in William Thesing, *Caverns of Night: Coal Mines in Art, Literature, and Film*, Columbia, SC: University of South Carolina Press, 2000, pp. 3–18.

49. Meikleham, *On the History and Art of Warming and Ventilating Rooms and Buildings*, II, p. 37.

50. François Delaporte, *Disease and Civilisation: the Cholera in Paris, 1832*, Cambridge, MA: The MIT Press, 1986, $_1$1832.

51. Henry Stephens, *Cholera: an Analysis of its Epidemic, Endemic, and Contagious Character, etc.*, London, 1849, pp. 1–2.

52. Charles Rosenberg, 'The Cause of Cholera: Aspects of Etiological Thought in 19th-Century America', in Leavitt and Numbers (eds), *Sickness and Health in America*, pp. 257–71.

53. As Rosenberg notes in the case of nineteenth-century America, the atmospheric theory had the added advantage of allowing one to sidestep the 'anti-social' aspects of a belief in contagion, which invariably necessitated practices of isolation and quarantine.

54. Rosenberg, 'The Cause of Cholera', pp. 261.

55. Charles Chapin, 'The End of the Filth Theory of Disease', in *Popular Science Monthly*, 60, 1902, pp. 234–9.

56. Quoted in John Grove, *Epidemics Examined and Explained: or, Living Germs Proved by Analogy to be a Source of Disease*, London, 1850, p. 16.

57. Nikolas Rose, *The Pyschological Complex: Psychology, Politics and Society in England 1869–1939*, London, Routledge and Kegan Paul, 1985, pp. 47–8.

58. Isaac Ray, *Mental Hygiene*, with a new Introduction by Frank Curran, New York and London: Hafner Publishing Company, 1968, $_1$1863, p. 66.

59. James Whorton, ' "Tempest in a Flesh-Pot": the Formulation of a Physiological Rationale for Vegetarianism', in Leavitt and Numbers, *Sickness and Health in America*, p. 326.

60. William Alcott, *Lectures on Life and Health; or, the Laws and Means of Physical Culture*, Boston, 1853, p. 226; cited in Whorton, ' "Tempest in a Flesh-Pot" ', p. 326.

61. Cassell and Co., *Cassell's Household Guide*, p. 338.

62. Molly Harrison, *The Kitchen in History*, New York: Charles Scribner's Sons, 1972, p. 124.

63. George Eliot (Mary Ann Cross, née Evans), *Scenes of Clerical Life*, London, 1858, p. 232, cited in Michael Waters, *The Garden in Victorian Literature*, Aldershot: Scolar Press, 1988, p. 50.

64. Harrison, *The Kitchen in History*, pp. 122–23.

65. Norman Shaftel, 'A History of the Purification of Milk in New York, or, "How Now, Brown Cow" ', in Leavitt and Numbers, *Sickness and Health in America*, p. 283. Even after the passage of the first Food and Drugs Act in 1860, country authorities in Great Britain were unable to control the quality of foodstuffs as there was much corruption and deceit in food-related industries. A subsequent law, passed in 1872, furthered the interests of those concerned with the purity of food by compelling the appointment of public analysis and shop inspectors.

66. Cassell and Co., *Cassell's Household Guide*, p. 257.

67. Bertrand Russell, *A History of Western Philosophy*, London: Unwin Hyman, 1984, $_1$1946, pp. 71–3.

68. Cassell and Co., *Cassell's Household Guide*, p. 99.

69. Cited in Shirley Murphy, *Our Homes and How to Make them Healthy*, London: Cassell, 1885, p. 2. The author is grateful to Ian Roderick for alerting him to Richardson's introductory comments in this book.

Chapter 5

1. Alistair Duckworth, 'Fiction and Some Uses of the Country House Setting from Richardson to Scott', *Landscape in the Gardens and the Literature of Eighteenth-Century England*, Los Angeles: The William Andrews Clark Memorial Library, 1981, p. 120.

2. Lynda Nead, *Victorian Babylon: People, Streets, and Images in Nineteenth-Century London*, New Haven, CT: Yale University Press, 2000, pp. 212–14.

3. George Perkins Marsh, *Man and Nature*, edited with an introduction by David Lowenthal, Cambridge MA: Harvard University Press, 1965, $_1$1864, p. 3.

4. Ian Hunter has argued that what we commonly celebrate as the 'freedom' of reading today is partly the result of the Victorians having established 'moral norms [that] would be realised through self-expressive techniques; and it was in this space that the forms of self-discovery organised around the individual would permit the realisation of new social norms at the level of the population'. See Hunter, *Culture and Government: the Emergence of Literary Education*, Basingstoke: Macmillan, 1988, p. 38.

5. As Joseph Caroll observed: 'British writers of the later nineteenth century lived in a long-cultivated, densely populated, and heavily industrialized country, but world exploration, colonial expansion, and the still fresh scientific revelations about geological time and evolutionary transformation offered a wide field for imaginative exploration of wild places.' See Caroll, 'The Ecology of Victorian Fiction', in *Philosophy and Literature*, 25:2, October, 2001, pp. 295–313, 305.

6. Robert Hunt, *Researches on Light: An Examination of All the Phenomena Connected with the Chemical and Molecular Changes Produced by the Influence of the Solar Rays*, London, 1844, p. 279.

7. John Hutton Balfour, *Outlines of Botany; Being an Introduction to the Study of the Structure, Functions, Classification, and Distribution of Plants*, Edinburgh, 1854, p. 522.

8. John Hutton Balfour, *Phyto-Theology; or, Botanical Sketches, intended to illustrate The Works of God in the Structure, Functions, and General Distribution of Plants*, London, 1851, p. 184.

9. Charles Smith, *Parks and Pleasure Grounds*, London, 1852, p. 171.

10. David Goodman, 'Fear of Circuses: Founding the National Museum of Victoria', in *Continuum*, 3:1, 1990, pp. 18–34.

11. Such principles could also be used to legitimize the divisions between human society. John Lindley's efforts to modernize botany were also aimed at distancing his profession from polite botany or 'amusement for the ladies'; see Ann Shteir, *Cultivating Women, Cultivating Science: Flora's Daughters and Botany in England, 1760–1860*, Baltimore, MD: The Johns Hopkins University Press, 1996.

12. Smith, *Parks and Pleasure Grounds*, p. 173.

13. Sue Minter, *The Greatest Glass House*, London: HMSO, 1990, p. 8.

14. John Mackenzie, 'Empire and the Ecological Apocalypse: the Historiography of the Imperial Environment', in Tom Griffiths and Libby Robin (eds), *Ecology and Empire: Environmental History of Settler Societies*, Melbourne: Melbourne University Press, 1997, pp. 215–28.

15. Balfour, *Phyto-Theology*, p. 13.

16. William Henry Hudson, *A Crystal Age*, London, 1887.

17. Karl Kroeber writes that Scott saw 'history as the evolution of competing styles of life' and that he was a 'topographical realist' suggesting the novelist's 'awareness of the stylistic uniqueness of past epochs and of geographically separated cultures'. See Kroeber, *Romantic Narrative Art*, Madison, WI: University of Wisconsin Press, 1960, p. 169; and Duckworth, 'Fiction and Some Uses of the Country House Setting', p. 111.

18. Joshua Major and Son, *The Ladies' Assistant in the Formation of Their Flower Gardens, etc.*, London, 1861, pp. 7–11.

19. Richard Grove, *Green Imperialism: Colonial Expansion, Tropical Island Edens, and the Origins of Environmentalism, 1600–1860*, Cambridge: Cambridge University Press, 1995.

20. Diana Loxley, *Problematic Shores: the Literature of Islands*, London: Macmillan, 1990, p. 3; Dorothy Lane, *The Island as Site of Resistance*, New York: Peter Lang, 1995, pp. 10–11.

21. Lane, *The Island as Site of Resistance*, p. 4.

22. Grove, *Green Imperialism*, p. 14.

23. Philip Gosse, *Wanderings Through the Conservatories at Kew*, London, 1857, p. 143.

24. Alexander von Humboldt, cited by Gosse, *Wanderings Through the Conservatories at Kew*, p. 144.

25. Major and Son, *The Ladies' Assistant*, pp. 7–11.

26. Denis Cosgrove, *Social Formation and Symbolic Landscape*, London: Croom Helm, 1984, Chapter 1.

27. Robin Evans, 'Figures, Doors and Passages', in *Architectural Design*, 4, April, 1978, pp. 267–78.

28. Of the many books detailing social aspects of nineteenth-century garden and home design practices, two particularly useful works are Joan Morgan and Alison Richards, *A Paradise Out of a Common Field*, New York: Harper and Row, 1990; and Asa Briggs, *Victorian Things*, London: Batsford, 1988.

29. David Thomson, *Handy Book of the Flower-Garden*, Edinburgh and London, 1868, p. 3.

30. Samuel Beeton, *The Book of Home Pets*, London, 1861, p. 803.

31. Cassell and Co., *Cassell's Household Guide*, London, 1869, p. 60.

32. Robson's comments are cited in Morgan and Richards, *A Paradise Out of a Common Field*, p. 25.

33. Cited by Morgan and Richards, *A Paradise Out of a Common Field*, p. 33.

34. Thomson, *Handy Book of the Flower-Garden*, pp. 276–85.

35. Ibid., pp. 281–2.

36. Cited by Morgan and Richards, *A Paradise Out of a Common Field*, p. 32.

37. David Hay, *The Laws of Harmonious Colouring Adapted to Interior Decorations with Observations on the Practice of House Painting*, 6th edn, Edinburgh, 1847, p. 42.

38. Robert Kerr, *The Gentleman's House*, 3rd edn, London, 1871, $_1$1864, p. 315.

39. Edward Kemp, *How to Lay Out a Small Garden*, London, 1850, p. 48.

40. Ibid., p. vii.

41. Ibid., p. 118.

42. Ibid., p. 49.

43. Hay, *The Laws of Harmonious Colouring*, pp. 8–9, 5.

44. Ibid., p. 67.

45. Mark Girouard, *Life in the English Country House*, New Haven, CT: Yale University Press, 1978, p. 238.

46. John Hix, *The Glasshouse*, London: Phaidon, 1996, $_1$1974, pp. 31, 93.

47. Stephen Ward, *Wardian Cases for Plants and their Application*, London, 1854, p. 22. Stephen Ward was the son of the inventor of the case, Nathaniel Ward.

48. Ibid., p. 25.

49. Nicholas Jardine, James Secord and Emma Spary (eds), *Cultures of Natural History*, Cambridge: Cambridge University Press, 1996, pp. 400–404.

50. Beeton, *The Book of Home Pets*, p. 683.

51. *Cassell's Household Guide*, p. 17.

52. Beeton, *The Book of Home Pets*, p. 770.

53. Ibid., p. 802.

54. *Cassell's Household Guide*, p. 65.

55. Kemp, *How to Lay Out a Small Garden*, p. 98.

56. Beeton, *The Book of Home Pets*, p. 771.

57. Morgan and Richards, *A Paradise Out of a Common Field*, pp. 35–9.

58. Thomson, *Handy Book of the Flower-Garden*, p. 3.

59. Hay, *The Laws of Harmonious Colouring*, p. 1.

60. Morgan and Richards, *A Paradise Out of a Common Field*, p. 39.

61. Cuthbert Johnson, *Our House and Garden: What We See, and What We Do Not See in Them*, London, 1864, p. 2.

62. William Robinson, *Hardy Flowers*, London, 1871, pp. 2–3.

63. John Ruskin (aka Kata Phusin), 'The Poetry of Architecture', in *The Architectural Magazine and Journal of Improvement in Architecture, Building, Furnishing, etc. Conducted by J.C. London*, November 1837.

64. John Ruskin, *The Poetry of Architecture; or, the Architecture of the Nations of Europe Considered in its Association with Natural Scenery and National Character*, Orpington, 1892.

65. Cited in John Illingworth, 'Ruskin and Gardening', in *Garden History*, 22:2, 1994, p. 22.

66. Ruskin, 'The Poetry of Architecture', in *Architectural Magazine*, November 1837.

67. Source: *Oxford English Dictionary* (original emphasis).

Chapter 6

1. Detailing the architecture of family life is the well-known work of Mark Girouard: *Life in the English Country House*, New Haven, CT: Yale University Press, 1978 and *The Victorian Country House*, New Haven, CT: Yale University Press, 1979.

2. Philip Gosse, *Wanderings Through the Conservatories at Kew*, London, 1857, p. 95.

3. Monica Greco, 'Psychosomatic Subjects and the "Duty to be Well": Personal Agency Within Medical Rationality', in *Economy and Society*, 22:3, August 1993.

4. Robert Kerr's *The Gentleman's House* was published in three editions: 1864, 1865 and 1871. A facsimile edition, published with an Introduction by Joseph Crook, was released in 1972 (New York: Johnson Reprint Company). Reactions to Kerr's book were mixed, though generally favourable, as Crook acknowledged. Many of the ideas Kerr expressed in the text were developed further and explained elsewhere, particularly in numerous articles and through his well-attended lectures. Kerr was a respected educator: co-founder and first President of the Architectural Association in London; second professor of the Arts and Construction at King's College and examiner and councillor at the RIBA.

5. Kerr, *The Gentleman's House*, reprint of 3rd edn (1871) with Introduction by Crook, New York: Johnson Reprint Company, 1972, pp. 45, 49. Subsequent references to Kerr's text are taken from his facsimile edn.

6. Joseph Crook, *The Dilemma of Style: Architectural Ideas from the Picturesque to the Post-Modern*, London: Murray, 1987.

7. John Summerson has depicted Kerr's search for a *raison d'être* behind the debate over style as an obsession in *Victorian Architecture: Four Studies in Evaluation*, New York: Columbia University Press, 1970.

8. Kerr, *The Gentleman's House*, 1972, pp. 50–51.

9. Ibid., pp. 58–9.

10. Ibid., p. 60.

11. Ibid., pp. 321–6.

12. Ibid., p. 315.

13. Ibid., pp. 315, 326.

14. Ibid., p. 332.

15. Robin Evans, 'Figures, Doors and Passages', in *Architectural Design*, 4, 1978, pp. 267–78.

16. Kerr, *The Gentleman's House*, 1972, pp. 313–17.

17. Ibid., pp. 330–33.

18. Ibid., pp. 314ff.

19. Ibid., p. 83.

20. Ibid., pp. 83–4.

21. Edward Kemp, *Landscape Gardening. How to lay out a garden … edited, revised and adapted to North America*, 4th edn, New York, 1911, ₁1850, p. 106.

22. Kerr, *The Gentleman's House*, 1972, pp. 70–71.

23. This is an implication of John Crowley's study, *The Invention of Comfort*, Baltimore, MD: The Johns Hopkins University Press, 2001.

24. Kemp, *Landscape Gardening*, p. 49.

25. Cassell and Co., *Cassell's Household Guide: Being a Complete Encyclopoedia of Domestic and Social Economy: and Forming a Guide to Every Department of Practical Life*, vol. 1, London, 1869, p. 99.

26. Evans, 'Figures, Doors and Passages', p. 270.

27. Esther Hewlett Copley, *Catechism of domestic economy*, London, 1851, pp. 17–18.

28. Nancy Armstrong, *Desire and Domestic Fiction*, Oxford: Oxford University Press, 1987, pp. 69, 75.

29. Armstrong writes: 'Their psychological differences made men political and women domestic rather than the other way around, and both therefore acquired identity on the basis of personal qualities that had formerly determined female nature alone.' As such, men were as much producers of domestic life as women governed it. Ibid., p. 4.

30. Kerr, *The Gentleman's House*, 1972, p. 74.

31. Robert Kerr, *A Small Country House*, London, 1873, p. 117.

32. David Thomson, *Handy Book of the Flower-Garden*, Edinburgh and London, 1868, p. 287.

33. Mary Cruger, *How She Did It or Comfort on $150 a Year*, New York, 1888, preface.

34. Ibid., p. 9.

35. Kerr, *A Small Country House*, p. 14.

36. Philippa Tristam, *Living Space in Fact and Fiction*, London: Routledge, 1989, p. 268.

37. Eliza Warren, *How I Managed My House on Two Hundred Pounds a Year*, London, 1864, p. v.

38. A longstanding literary device, character became a subject of considerable reflection in nineteenth-century thought, serving as a 'projection or correlate of the reader's moral self and personality'. Ian Hunter, 'Reading Character', in *Southern Review*, 16:2, July 1983, pp. 230–34.

39. Kerr, *The Gentleman's House*, 1972, pp. 69–70.

40. Kerr, *A Small Country House*, p. 11.

41. Kerr, *The Gentleman's House*, 1972, p. 69.

42. John Claudius Loudon, *The Suburban Gardener and Villa Companion*, London, 1838, p. 325. The emphasis in the passage is mine.

43. Melanie Louise Simo, *Loudon and the Landscape*, New Haven, CT: Yale University Press, 1988, p. 263.

44. Kerr, *The Gentleman's House*, 1972, p. 69.

45. This re-alignment of relations between designer and client parallels shifts which occurred late in the eighteenth century between doctors and their patients. This marked a significant alignment of biological and medical interests. In broad social terms, the consequence of such a development for the architectural profession was perhaps not as great as it was for doctors given the prestige afforded to health professionals today. Still, one could argue that the use of characterization and the attendant design practices were important in defining the parameters of professional architectural practice and landscape architecture. See Nikolas Rose, *The Psychological Complex: Psychology, Politics and Society in England, 1869–1938*; London: Routledge, 1985; Michel Foucault, *The Birth of the Clinic*, London: Tavistock Press, 1973, ₁1963, Chapter 6.

46. Anthony Vidler, 'Psychopathologies of Modern Space: Metropolitan Fear from Agoraphobia to Estrangement', in Michael Roth (ed.), *Rediscovering History: Culture, Politics, and the Psyche*, Stanford, CA: Stanford University Press, 1994.

47. Greco, 'Psychosomatic Subjects', pp. 359–60.

48. Ibid., p. 369.

49. It was a place where, in Grahame Burchell's terms, government mediated between powers of domination and techniques of the self. See Burchell, 'Liberal government and the techniques of the self' in *Economy and Society*, 22:3, August 1993, p. 268.

Chapter 7

1. See Stephen Kern, *The Culture of Time and Space 1880–1918*, Cambridge, MA: Harvard University Press, 1983.

2. Norbert Elias, *Time: An Essay*, Oxford: Blackwell, 1992, p. 21.

3. Ferdinand Braudel, *The Structures of Everyday Life*, London: Collins, 1981, pp. 72, 483ff.

4. Eric Lampard, 'The Urbanizing World', in Harold Dyos and Michael Wolff (eds), *The Victorian City: Images and Reality*, vol. 1, London: Routledge, 1976, pp. 19–23.

5. John Strang, *Necropolis Glasguensis; With Observations on Ancient and Modern Tombs and Sepulture*, Glasgow, 1831, pp. 28–9.

6. George Collison, *Cemetery Interment: Containing a Concise History of the Modes of Interment Practised by the Ancients; Descriptions of Père-Lachaise, the Eastern Cemeteries, and those of America; the English Metropolitan and Provincial Cemeteries, and More Particularly of the Abney Park Cemetery, at Stoke Newington*, London, 1840.

7. Richard Etlin, *The Architecture of Death: the Transformation of the Cemetery in Eighteenth-Century Paris*, Cambridge, MA: The MIT Press, 1984, p. 359.

8. Stanley French, 'The Cemetery as Cultural Institution: the Establishment of Mount Auburn and the "Rural Cemetery" Movement', in David Stannard (ed.), *Death in America*, Pennsylvania: University of Pennsylvania Press, 1975, pp. 69–70.

9. Mary Douglas (1992), cited in Ian Hacking, 'Memory Sciences, Memory Politics', in Paul Antze and Michael Lambek (eds), *Tense Past: Cultural Essays in Trauma and Memory*, London: Routledge, 1996.

10. Harriet Martineau writing in 1838, cited in French, 'The Cemetery as Cultural Institution', pp. 86–7.

11. Cited in John Claudius Loudon, *An Encyclopaedia of Gardening*, I, London, 1835, ₁1822, p. 111.

12. Ibid., p. 111.

13. Ibid., p. 112.

14. John Claudius Loudon, *Parochial Institutions; or, An Outline of a Plan for a National Education Establishment, Suitable to the Children of All Ranks, from Infancy to the Age of Puberty; as a Substitute for the National Churches of England, Scotland, and Ireland*, London, 1829, p. 2.

15. Ibid., p. 5.

16. John Claudius Loudon, *On the Laying Out, Planting, and Managing of Cemeteries; and on the Improvement of Churchyards*, London, 1843, pp. 12–13.

17. Paul Coones, 'Kensal Green Cemetery', in *Ancient Monuments Society Transactions*, 31, 1987, p. 49.

18. James Steven Curl, 'Nunhead Cemetery, London; History of the Planning, Architecture, Landscaping and Fortunes of a Great Nineteenth-Century Cemetery', in *Ancient Monuments Society Transactions*, 22 (1977), p. 31.

19. John Richards, *Essay on Cemetery Interments, Awarded the Prize Offered by the Directors of the Reading Cemetery Company. Edited with the Report of the Select Committee on the Health of Towns and Selections from the Evidence Taken Before the Committee*, London, 1843, p. 10.

20. David Stow, *The Training System, the Moral Training School, and the Normal Seminary*, London, 1850. Joseph Lancaster was to give his name to the system of 'Lancastrian' schools; the full title of his most influential treatise is *Improvements in Education as it Respects the Industrious Classes of the Community* (London, 1838).

21. Loudon, *Parochial Institutions*, p. 6.

22. Melanie Louise Simo, *Loudon and the Landscape*, New Haven, CT: Yale University Press, 1988, p. 197.

23. Ian Hunter, *Culture and Government: the Emergence of Literary Education*, Basingstoke: Macmillan, 1988, p. 5.

24. James Stevens Curl, 'Young's *Night Thoughts* and the Origins of the Garden Cemetery', in *Journal of Garden History*, 14, April/June 1994. For Curl's study of Loudon and the garden cemetery, see *A Celebration of Death*, London: Constable, 1980, Chapter 8.

25. Cited in Curl, 'Young's *Night Thoughts*', p. 93.

26. Strang, cited in Curl, 'Young's *Night Thoughts*', p. 92. This author's emphasis.

27. Strang, *Necropolis Glasguensis*, p. 29.

28. Richards, *Essay on Cemetery Interments*, p. 13.

29. Strang, *Necropolis Glasguensis*, pp. 32–3.

30. Samuel Smiles, *Self Help; With Illustrations of Character and Conduct*, London, 1859, p. 23.

31. Loudon, *Parochial Institutions*, p. 4.

32. Ibid., p. 5.

33. Richards, *Essay on Cemetery Interments*, p. 5.

34. Strang, *Necropolis Glasguensis*, pp. 6–7.

35. Richards, *Essay on Cemetery Interments*, p. 13.

36. Collison, *Cemetery Interment*, p. 303.

37. Ibid., p. 306.

38. Loudon, *On the Laying Out, Planting, and Managing of Cemeteries*, p. 8.

39. Richards, *Essay on Cemetery Interments*, p. 13.

40. Cited in Collison, *Cemetery Interment*, p. 11.

41. Margaretta J. Darnall, 'The American Cemetery as Picturesque Landscape', in *Winterthur Portfolio*, 18:4, Winter 1983, p. 250.

42. Strang, *Necropolis Glasguensis*, p. 61.

43. Ibid., p. 47.

44. Loudon, *On the Laying Out, Planting, and Managing of Cemeteries*, p. 74.

45. Strang, *Necropolis Glasguensis*, p. 46.

46. Loudon, *On the Laying Out, Planting, and Managing of Cemeteries*, p. 69.

47. Ibid., p. 97.

48. Ibid., pp. 12–13.

49. Ibid., p. 78.

50. Humphry Repton had argued as much when advising that 'if that road be greatly circuitous, no one will use it'. See Repton, *An Enquiry into the Changes of Taste in Landscape Gardening* (London, 1806), cited in Darnall, 'The American Cemetery as Picturesque Landscape', p. 264.

51. Strang, *Necropolis Glasguensis*, p. 61.

52. Ibid., pp. 36–7.

53. Curl noted, however, the horror expressed by many travellers of the disregard for Protestant remains in the burial practices of Catholic nations. This imparted an even more forlorn quality to the melancholic musings of visitors to such Protestant cemeteries as that in Rome.

54. Comments recorded in Collison, *Cemetery Interment*, p. 296.

55. For this citation and other references to the importance of restraint as regards the trappings of death see Pat Jalland, *Death in the Victorian Family*, Oxford: Oxford University Press, 1996, pp. 194–209, 292.

56. Such meanings are listed in Hugh Meller, *London Cemeteries, an Illustrated Guide and Gazetteer*, Amersham: Avesbury Publishing Company, 1985, and cited in Coones, 'Kensal Green Cemetery', p. 58.

57. Cited in Curl, 'Nunhead Cemetery', p. 33.

58. Paul Joyce, *A Guide to Abney Park Cemetery*, London, 1994, p. 14.

59. Philippe Aries, *The Hour of Our Death*, New York: Vintage Books, 1982, ₁1977, p. 580.

60. Joyce, *A Guide to Abney Park Cemetery*, p. 66.

61. Philippe Aries, *Centuries of Childhood: a Social History of Family Life*, trans. from French by Robert Baldick, New York: Vintage Books/Random House; London: Jonathan Cape, 1962, ₁1960.

62. Declan Kiberd, 'Yeats, Childhood and Exile', in Paul Hyland and Neil Sammells (eds), *Irish Writing: Exile and Subversion*, London: Macmillan, 1991, p. 127.

63. Edward Kemp, *How to Lay Out a Small Garden*, London, 1850, p.103.

64. Laurence Kirmayer, 'Landscapes of Memory: Trauma, Narrative, and Dissociation', in Antze and Lambek, *Tense Past*, p. 179.

Conclusion

1. Immanuel Kant, 'Idea for a Universal History from a Cosmopolitan Point of View' (1784), cited in Ian Hacking, *The Taming of Chance*, Cambridge: Cambridge University Press, 1990, p. 15.

2. Francis Galton, *Hereditary Genius*, London: Macmillan, 1869.

3. Hacking, *The Taming of Chance*, p. 38.

4. Nikolas Rose, *The Psychological Complex: Psychology, Politics, and Society in England, 1869–1939*, London: Routledge, 1985, p. 73.

5. Paul Hirst and Penny Woolley, 'Nature and Culture in Social Science: the Demarcation of Domains of Being in Eighteenth Century and Modern Discourses', in *Geoforum*, 16:2, 1985, p. 153.

6. Roger Shattuck, *The Forbidden Experiment: the Story of the Wild Boy of Aveyron*, New York: Farrar, Straus and Giroux, 1980.

7. Rose, *The Psychological Complex*, pp. 16–17.

8. John Claudius Loudon, *An Encyclopaedia of Gardening*, London, 1835, ₁1822, p. 112.

9. Addressing a similar point, John Dixon Hunt described how literary genres in the eighteenth century were linked, 'rather uneasily', to new possibilities for landscape gardening in the nineteenth century. Viewers brought to familiar garden scenery composed of emblematic statuary and architectural forms new modes of interpretation and expectations of meaning, while responses to the natural world became more malleable, personal and introspective. See Hunt, *The Figure in the Landscape: Poetry, Painting, and Gardening During the Eighteenth Century*, Baltimore, MD: Johns Hopkins University Press, 1976, pp. 68, 196.

10. Michel Foucault, *The Order of Things*, New York: Vintage Books Edition, 1973, ₁1966, p. 229.

11. Michael Waters, *The Garden in Victorian Literature*, Aldershot: Scolar Press, 1988, p. 51.

12. Ibid., p. 53.

13. George Levine, *Darwin and the Novelists: Patterns of Science in Victorian Fiction*, Cambridge, MA: Harvard University Press, 1988, pp. 34–49.

14. Cassell and Co., *Cassell's Household Guide: Being a Complete Encyclopoedia of Domestic and Social Economy: and Forming a Guide to Every Department of Practical Life*, vol. I, London, 1869, p. 338.

15. Ibid., p. 312.

16. Asa Briggs, *Victorian Things*, London: Batsford, 1988.

17. James Cameron, *Is My House Healthy? How to Find Out*, London, 1892, p. 1.

18. Levine, *Darwin and the Novelists*, p. 47.

19. Cuthbert Johnson, *Our House and Garden: What We See, and What We Do Not See in Them*, London, 1864, p. 147.

20. Charles Darwin, 'The Formation of Vegetable Mould' (1881), reproduced in Albert Howard, *Darwin on Humus and the Earthworm*, London: Faber and Faber, 1945, pp. 1 , 65.

21. Ibid., p. 145.

22. Ibid., p. 92.

23. Levine, *Darwin and the Novelists*, p. 16.

24. John Hutton Balfour, *Phyto-Theology; or, Botanical Sketches, Intended to Illustrate the Works of God in the Structure, Functions, and General Distribution of Plants*, London, 1851, p. 13.

25. David Thomson, *Handy Book of the Flower-Garden*, Edinburgh and London, 1868, p. 1.

26. The Eden Project website: http://www.edenproject.com.

27. Recorded in *Knight's Cyclopaedia of London* (1851) and cited in John Hix, *The Glasshouse*, London: Phaidon, 1996, ₁1974, p. 122, and in Joan Morgan and Alison Richards, *A Paradise Out of a Common Field*, New York: Harper and Row, 1990, back cover.

28. Comments of one 11-year-old visitor, in *The Guardian*, 4 March 2002.

29. Mark Townsend, 'Car Fumes Blight Eden's Green Vision', in *The Observer*, 2 June 2002.

30. Alexander Welsh, 'Dome Decoy', *The Guardian*, 3 July 2002.

31. Interview with Tim Smit, in *The Observer*, 31 March 2002.

32. George Hersey, *The Monumental Impulse*, Cambridge, MA: The MIT Press, 1999; Nancy Aiken, *The Biological Origins of Art*, Westport, CT: Praeger, 1998; and Jay Appleton, 'Pleasure and the Perception of Habitat', in Barry Sadler and Allen Carlson (eds), *Environmental Aesthetics: Essays in Interpretation*, Victoria, BC: University of Victoria, 1982; and 'Prospects and Refuges Re-Visited', in *Landscape Journal*, 3:2, 1984.

33. An assessment of evolutionary psychology is found in Hilary and Steven Rose (eds), *Alas, Poor Darwin: Arguments Against Evolutionary Psychology*, London: Vintage, 2001.

34. Jack Rostron, *Sick Building Syndrome: Concepts, Issues and Practice*, London: Spon, 1997; Nadav Malin, 'Indoor Air Quality: Some Old and Some New Sick-Building Culprits', in *Architectural Record*, 184:2, February 1996, and D. M. Rowe, 'Sick Building Syndrome: the Mystery and the Reality', in *Architectural Science Review*, 37, September 1994.

Bibliography

First publication date is indicated by a subscript '$_1$'.

Historical sources

Agricola, Georgius, *De Re Metallica* (Basel, 1556)

Alcott, William, *Lectures on Life and Health; or, the Laws and Means of Physical Culture* (Boston, 1853)

Anon., 'Fogs in a Metropolitan Light', *The Builder*, 23:1173 (29 July 1865), 537–8

—— *The House and its Surroundings* (London, 1878)

Ansted, David, *Natural History of the Inanimate Creation: Being a Guide to the Scenery of the Heavens, the Phenomena of the Atmosphere, the Structure and Geological Features of the Earth, and its Botanical Productions* (London, 1856)

Balfour, John Hutton, *Phyto-Theology; or, Botanical Sketches, Intended to Illustrate the Works of God in the Structure, Functions, and General Distribution of Plants* (London, 1851)

—— *Outlines of Botany; Being an Introduction to the Study of the Structure, Functions, Classification, and Distribution of Plants* (Edinburgh, 1854)

Beeton, Samuel, *The Book of Home Pets* (London, 1861)

Bischof, Gustav, 'On Putrescent Organic Matter in Potable Water', *Proceedings of the Royal Society of London*, 26 (London, 1877)

Britten, James and Holland, Robert, *A Dictionary of English Plant-Names* (London, 1886)

Burroughs, Edgar Rice, *Tarzan of the Apes* (New York, 1912)

Cameron, James, *Is My House Healthy? How to Find Out* (Leeds, 1892)

Campbell, Colen, *Vitruvius Britannicus* (London, 1715–25)

Cassell and Co., *Cassell's Household Guide: Being a Complete Encyclopoedia of Domestic and Social Economy: and Forming a Guide to Every Department of Practical Life* (London, 1869)

Coleridge, Samuel, *General Introduction to the Encyclopaedia Metropolitana; or, A Preliminary Treatise on Method* (London, 1818)

Collison, George, *Cemetery Interment: Containiming a Concise History of the Modes of Interment Practised by the Ancients; Descriptions of Père-Lachaise, the Eastern Cemeteries, and those of America; the English Metropolitan and Provincial Cemeteries, and More Particularly of the Abney Park Cemetery, at Stoke Newington* (London, 1840)

Copley, Esther Hewlett, *Catechism of Domestic Economy* (London, 1851)

Cruger, Mary, *How She Did It or Comfort on $150 a Year* (New York, 1888)

Darwin, Charles, *Journal of researches into the natural history and geology of the countries visited during the voyage of H.M.S. 'Beagle' round the world*, 10th edn (London, 1891)

Darwin, Charles, 'The Formation of Vegetable Mould' (1881); reproduced in Albert Howard, *Darwin on Humus and the Earthworm* (London: Faber and Faber, 1945)

Darwin, Charles Robert, *On the Origin of Species by Means of Natural Selection, or the Preservation of Favoured Races in the Struggle for Life* (London, 1859)

Davy, Christopher, *The Architect, Engineer and Operative Builder's Constructive Manual, or, A Practical and Scientific Treatise on the Construction of Artificial Foundations for Buildings, Railways, &c.* (London, 1839)

Davy, John, *Lectures on the Study of Chemistry, in Connexion with the Atmosphere, the Earth and the Ocean, etc.* (London, 1849)

Davy, Humphry, *Elements of Agricultural Chemistry* (London, 1813)

Doré, August, *London: A Pilgrimage* (London, 1872)

Downing, Andrew, *A Treatise on the Theory and Practice of Landscape Gardening Adapted to North America, With a View to the Improvement of County Residences*, 6th edn (New York, 1859)

Eliot, George, *Scenes of Clerical Life* (Edinburgh and London, 1858)

Fergusson, James, *A History of Architecture in All Countries from the Earliest Times to the Present Day*, 3rd edn (London, 1893)

Galton, Sir Francis, *Hereditary Genius* (London, 1869)

Girdlestone, Edward, *Our Debt and Duty to the Soil: or, The Poetry and Philosophy of Sewage Utilization* (London, 1878)

Gosse, Philip, *Wanderings Through the Conservatories at Kew* (London, 1857)

Grove, John, *Epidemics Examined and Explained: or, Living Germs Proved by Analogy to be a Source of Disease* (London, 1850)

Hales, Stephen, *Vegetable Staticks, or an Account of Some Statical Experiments on the Sap in Vegetables* (London, 1727)

—— *A Description of Ventilators: Whereby Great Quantities of Fresh Air May with Ease be Conveyed into Mines, Gaols, Hospitals, Work-Houses and Ships, in Exchange for their Noxious Air, etc.* (London, 1743)

Harris, John, 'Some Microscopical Observations of Vast Numbers of Animalcula Seen in Water, *Philosophical Transactions*, 19 (London, 1695–7)

Hay, David, *The Laws of Harmonious Colouring Adapted to Interior Decorations with Observations on the Practice of House Painting*, 6th edn (Edinburgh, 1847)

Hill, John, 'The Sleep of Plants' (1757), *Botanical Tracts* (London, 1762)

Home, Henry, *Gentleman Farmer* (Edinburgh, 1776)

Hooker, Joseph Dalton, *Report on the Progress and Condition of the Royal Gardens at Kew, During the Year 1866* (London, 1866)

Hudson, William Henry, *A Crystal Age* (London, 1887)

Hume, David, *Dialogues Concerning Natural Religion* (London, 1779, $_1$1751)

Hunt, Robert, *Researches on Light: An Examination of All the Phenomena Connected with the Chemical and Molecular Changes Produced by the Influence of the Solar Rays* (London, 1844)

Ingenhousz, Jan, *Experiments upon Vegetables, Discovering their Great Power of Purifying the Common Air in Sunshine, and of Injuring it in the Shade and at Night*, London, 1779.

Jefferies, William, *After London, or, Wild England* (London, 1885)

Johnson, Cuthbert, *The Objects and History of the Thames Improvement Company* (London, 1839)

—— *On the Cottages of Agricultural Labourers: with Economical Working Plans, and Estimates for their Improved Construction* (London, 1847)

—— *Our House and Garden: What We See, and What We Do Not See In Them*, London, 1864

Jones, Owen, *Grammar of Ornament* (London, 1856)

Kant, Immanuel, *Critique of Pure Reason*, translated by J. Bernard (New York: Hafner Publishing Company, 1965, $_1$1781)

—— 'Idea for a Univeral History from a Cosmopolitan Point of View', cited in
    Hacking, Ian, *The Taming of Chance* (Cambridge: Cambridge University Press,
    1990, ₁1784)
Kearley, George , *Links in the Chain; or Popular Chapters on the Curiosities of Animal Life*
    (London, 1862)
Kemp, Edward, *How to Lay Out a Small Garden* (London, 1850)
—— *Landscape Gardening or How to Lay Out a Garden … Edited, Revised and Adapted to
    North America*, 4th edn (New York, 1911, ₁1850)
Kerr, Robert, *A Small Country House* (London, 1873)
Kerr, Robert, *The Gentleman's House*, 3rd edn, London, 1871, ₁1864
—— *The Gentleman's House*, reprint of 3rd edn (1871) with Introduction by Joseph
    Crook (New York: Johnson Reprint Company, 1972)
Knight, Thomas, *The Report of a Committee of the Horticultural Society of London*
    (London, 1805)
—— 'On the Ill Effects of Excessive Heat in Forcing-Houses During the Night'
    (1814), in *Transactions of the Horticultural Society of London*, II (London, 1817)
Lackington, James, *Lives of the Alchemists* (London, 1815)
Lancaster, Joseph, *Improvements in Education as it Respects the Industrious Classes of the
    Community* (London, 1838)
Laugier, Marc-Antoine, *An Essay on Architecture*, trans. with Introduction by
    Wolfgang and Anni Hermann (Los Angeles: Hennessey and Ingalls, 1977, ₁1753)
Locke, John, *An Essay Concerning Human Understanding* (London, 1748, ₁1690)
Loudon, John Claudius, *A Short Treatise on Several Improvements, Recently Made in Hot-
    Houses* (Edinburgh, 1805)
—— *Remarks on the Construction of Hothouses* (London, 1817)
—— *An Encyclopaedia of Gardening* (London, 1835, ₁1822)
—— *Parochial Institutions; or, An Outline of a Plan for a National Education Establishment,
    Suitable to the Children of All Ranks, from Infancy to the Age of Puberty; as a Substitute
    for the National Churches of England, Scotland, and Ireland* (London, 1829)
—— *The Suburban Gardener and Villa Companion* (London, 1838)
—— *On the Laying Out, Planting, and Managing of Cemeteries; and on the Improvement of
    Churchyards* (London, 1843)
Macartney, Lord George, *An Embassy to China: Being the Journal Kept by Lord Macartney
    During his Embassy to the Emperor Ch'ien-lung 1793–1794* (1797), edited with an
    Introduction and Notes by Launcelot Cranmer-Byng (London: Longman, 1962)
Mackenzie, George, 'Description of an Economical Hot-House' (1814), *Memoires of the
    Caledonian Horticultural Society*, II (Edinburgh, 1818)
—— 'On the Form which the Glass of a Forcing-House Ought to Have, in Order to
    Receive the Greatest Possible Quantity of Rays from the Sun' (1815), in
    Transactions of the Horticultural Society of London, II (London, 1817)
Major, Joshua and Son, *The Ladies' Assistant in the Formation of their Flower Gardens,
    etc.* (London, 1861)
Malthus, Thomas Robert, *An Essay on the Principle of Population, as it Affects the Future
    Improvement of Society; with Remarks on the Speculations of W. Godwin, M. Condorcet
    and Other Writers* (London, 1798)
Marsh, George Perkins, *Man and Nature; or, Physical Geography as Modified by Human
    Action*, edited with an introduction by David Lowenthal (Cambridge, MA:
    Harvard University Press, 1965, London, ₁1864)
Meikleham, Robert, *On the History and Art of Warming and Ventilating Rooms and
    Buildings, etc.* (London, 1845)
Muthesius, Hermann, *The English House* (Berlin, 1908)
Nieuhof, Jan, *An Embassy from the East India Company of the United Provinces to the
    Grand Tartar Cham, Emperor of China*, trans. from the Latin edn by John Ogilby
    (London, 1669)

Papworth, John, *Rural Residences: Consisting of a Series of Designs for Cottages, Decorated Cottages, Small Villas, and other Ornamental Buildings* (London, 1818)

Phillips, Samuel, *Guide to the Crystal Palace and its Park and Gardens* (London, 1858)

Prior, Richard, *On the Popular Names of British Plants, Being an Explanation of the Origin and Meaning of the Names of Our Indigenous and Most Commonly Cultivated Species* (London, 1870)

Repton, Humphry, *An Enquiry into the Changes of Taste in Landscape Gardening* (London, 1806)

—— *Fragments on the Theory of Landscape Gardening* (London, 1816)

Richards, John, *Essay on Cemetery Interments, Awarded the Prize Offered by the Directors of the Reading Cemetery Company. Edited with the Report of the Select Committee on the Health of Towns and Selections from the Evidence Taken Before the Committee* (London, 1843)

Robinson, William, *Hardy Flowers* (London, 1871)

—— *The Wild Garden*, 4th edn (London, 1894, reprint 1977)

Romanes, George, 'Animal Intelligence', in *Fortnightly Review* (December 1881)

Rousseau, Jean-Jacques, *Discourse on the Origin of Inequality Among Men* ($_1$1755) and *The Social Contract* ($_1$1762), cited in George Cole (trans.), *The Social Contract, and Discourses* (London: Dent, 1986)

Ruskin, John, *Architectural Magazine and Journal of Improvement in Architecture, Building, Furnishing, etc. Conducted by J.C. Loudon* (November 1837)

—— *The Poetry of Architecture; or, the Architecture of the Nations of Europe Considered in its Association with Natural Scenery and National Character* (Orpington, 1892)

—— *The Seven Lamps of Architecture* (London, 1849)

—— *The Stones of Venice*, 3 vols (London, 1851–53)

Schlegel, Friedrich von, *The Philosophy of Life and Philosophy of Language; in a course of lectures*, translated from the German by A.J.W. Morrison (Bonn, 1848, $_1$1828)

Semper, Gottfried, 'The Four Elements of Architecture', in *The Four Elements of Architecture and Other Writings* (1815), trans. and with an Introduction by Harry Mallgrave and Wolfgang Hermann (Cambridge: Cambridge University Press, 1989)

Smiles, Samuel, *Self Help; With Illustrations of Character and Conduct* (London, 1859)

Smith, Charles, *Parks and Pleasure Grounds* (London, 1852)

Smith, John, *History of the Royal Gardens at Kew* (Royal Botanic Gardens, Kew: *Kewensia* Collection, 1880)

Stephens, Henry, *Cholera: an Analysis of its Epidemic, Endemic, and Contagious Character, etc.* (London, 1849)

Stevenson, John James, *House Architecture* (London, 1880)

Stow, David, *The Training System, the Moral Training School, and the Normal Seminary* (London, 1850)

Strang, John, *Necropolis Glasguensis; With Observations on Ancient and Modern Tombs and Sepulture* (Glasgow, 1831)

Swift, Jonathan, *Travels into Several Remote Nations of the World* (London, 1726)

Thackrah, Charles, *The Effects of the Principal Arts, Trades, and Professions, and of Civic States and Habits of Living, on Health and Longevity, etc.* (London, 1831)

Thompson, Robert, *The Gardener's Assistant* (London, 1881)

Thomson, Henry, *The Choice of a Profession* (London, 1857)

Thomson, David, *Handy Book of the Flower-Garden* (Edinburgh and London, 1868)

Tod, George, *Plans, Elevations and Sections of Hothouses, Green-Houses, An Aquarium, Conservatories, &c* (London, 1812)

Ward, Nathanial, *On the Growth of Plants in Closely Glazed Cases*, 2nd edn (London, 1852)

Ward, Stephen, *Wardian Cases for Plants and their Application* (London, 1854)

Warren, Eliza, *How I Managed My House on Two Hundred Pounds a Year* (London, 1864)

Wilkinson, George, *Practical Geology and Ancient Architecture* (Dublin, 1845)

Williams, John, 'On the Cultivation of the Vine in Forcing-Houses, with Observations on Forcing Peaches' (1814), in *Transactions of the Horticultural Society of London*, II (London, 1814)

Wollaston, William, 'On a Periscopic Camera Obscura and the Microscope', in *Philosophical Transactions of the Royal Society of London*, 102 (London, 1812)

Wood, John, *The Illustrated Natural History* (London, 1863)

Contemporary sources

Abbot, Barbara, 'Water = H_2O', *Mind*, 108:429 (1999), 145–8

Ackroyd, Peter, *London, the Biography* (London: Chatto and Windus, 2000)

Adams, Annemarie, *Architecture in the Family Way* (Montreal: McGill–Queen's University Press, 1996)

Aiken, Nancy, *The Biological Origins of Art* (Westport, CT: Praeger, 1998)

Allan, Mea, *The Hookers of Kew, 1785–1911* (London: Michael Joseph/Folio Society, 1967)

Antze, Paul and Lambek, Michael (eds), *Tense Past: Cultural Essays in Trauma and Memory* (London: Routledge, 1996)

Appleton, Jay, 'Pleasure and the Perception of Habitat', in Sadler, Barry and Carlson, Allen (eds), *Environmental Aesthetics: Essays in Interpretation* (Victoria, BC: University of Victoria, 1982)

—— 'Prospects and Refuges Re-Visited', in *Landscape Journal*, 3:2 (1984)

Archer, John, *The Literature of British Domestic Architecture, 1715–1842* (Cambridge, MA: The MIT Press, 1985)

Aries, Philippe, *The Hour of Our Death* (New York: Vintage Books, 1982, $_1$1977)

—— *Centuries of Childhood: a Social History of Family Life*, trans. from French by Robert Baldick (New York: Vintage Books/Random House; London: Jonathan Cape, 1962, $_1$1960)

Armstrong, Nancy, *Desire and Domestic Fiction* (Oxford: Oxford University Press, 1987)

Banham, Reyner, *The Architecture of the Well-Tempered Environment* (London: Architectural Press, 1969)

Bartra, Roger, *Wild Men in the Looking Glass: the Mythic Origins of European Otherness* (Ann Arbor, MI: University of Michigan Press, 1994)

Behne, Adolf, *The Modern Functional Building* (Santa Monica, CA: Getty Research Institute for the History of Art and the Humanities, 1996)

Bevis, Phil, Cohen, Michele and Kendall, Gavin, 'Archaeologizing Genealogy: Michel Foucault and the Economy of Austerity', in Gane, Mike and Johnson, Terry (eds), *Foucault's New Domains* (London: Routledge, 1993)

Bingham, Madeleine, *The Making of Kew* (London: Michael Joseph/Folio Society, 1975)

Bourdon, David, *Designing the Earth: the Human Impulse to Shape Nature* (New York: Abrams, 1995)

Bowler, Peter, *The Fontana History of the Environmental Sciences* (London: Fontana Press, 1992)

Braudel, Ferdinand, *The Structures of Everyday Life* (London: Collins, 1981)

Brazell, John H., *London Weather* (London: HMSO, 1968)

Briggs, Asa, *Iron Bridge to Crystal Palace: Impact and Images of the Industrial Revolution* (London: Thames and Hudson, 1979)

—— *Victorian Things* (London: Batsford, 1988)

Brown, Beverley and Cousins, Mark, 'The Linguistic Fault: the Case of Foucault's Archaeology', in *Economy and Society*, 9:3 (August, 1980)

Buchanan, Robert, *The Engineers: a History of the Engineering Profession in Britain, 1750–1914* (London: Jessica Kingsley, 1989)

Bullock, Alan and Stallybrass, Oliver (eds), *The Fontana Dictionary of Modern Thought* (London: Fontana, 1983, ₁1977)

Burchell, Grahame, *Economy and Society*, 22:3 (August 1993)

Burroughs, Edgar Rice, *Tarzan of the Apes* (New York, 1912)

Canguilhem, Georges, *The Normal and the Pathological* (New York: Zone Books, 1989, ₁1943)

—— *Ideology and Rationality in the History of the Life Sciences* (Cambridge, MA: The MIT Press, 1988)

Caroll, Joseph, 'The Ecology of Victorian Fiction', in *Philosophy and Literature*, 25:2 (October, 2001)

Cartwright, Thomas, *Domestic Science. The Science of Domestic Economy and Hygiene Treated Experimentally* (London, 1900)

Cassedy, James, 'The Flamboyant Colonel Waring: an Anticontagionist Holds the American Stage in the Age of Pasteur and Koch', in Judith Leavitt and Ronald Numbers (eds), *Sickness and Health in America: Readings in the History of Medicine and Public Health* (Madison, WI: Wisconsin University Press, 1978)

Chadwick, George, *Works of Sir Joseph Paxton* (London: Architectural Press, 1961)

Chapin, Charles, 'The End of the Filth Theory of Disease', *Popular Science Monthly*, 60 (1902), 234–9

Cipolla, Carlo (ed.), *The Emergence of Industrial Societies* (London: Fontana, 1973)

Colquhoun, Alan, *Modernity and the Classical Tradition: Architectural Essays 1980–1987* (Cambridge, MA: The MIT Press, 1989)

Conner, Patrick, *Oriental Architecture in the West* (London: Thames and Hudson, 1979)

Conway, Hazel, *People's Parks: the Development and Design of Victorian Parks* (Cambridge: Cambridge University Press, 1991)

Cook, Ernest, *Introduction to John Ruskin, 'The Stones of Venice'* (London: George Allen, 1904)

Coones, Paul, 'Kensal Green Cemetery', in *Ancient Monuments Society Transactions*, 31 (1987)

Corbin, Alain, *The Foul and the Fragrant: Odor and the French Social Imagination* (Leamington Spa: Berg, 1986)

Cosgrove, Denis, *Social Formation and Symbolic Landscape* (London: Croom Helm, 1984)

Cousins, Mark and Hussain, Athar, *Michel Foucault: Theoretical Traditions in the Social Sciences* (London: Macmillan, 1984)

Crary, Jonathan, *Techniques of the Observer: on Vision and Modernity in the Nineteenth Century* (Cambridge, MA: The MIT Press, 1990)

Crook, Joseph, *The dilemma of style: architectural ideas from the picturesque to the postmodern* (London: Murray, 1987)

Crowley, John, *The Invention of Comfort* (Baltimore, MD: The Johns Hopkins University Press, 2001)

Curl, James Stevens, 'Nunhead Cemetery, London; a History of the Planning, Architecture, Landscaping and Fortunes of a Great Nineteenth-Century Cemetery', in *Ancient Monuments Society Transactions*, 22 (1977)

—— *A Celebration of Death* (London: Constable, 1980)

—— 'Young's *Night Thoughts* and the Origins of the Garden Cemetery', in *Journal of Garden History*, 14 (April/June 1994)

Dallimore, William, 'A Gardener's Reminiscences' (Royal Botanic Garden, Kew: Kewensia Collection, unpublished manuscript, *c.* 1955)

Danahay, Martin, 'The Aesthetics of Coal: Representing Soot, Dust, and Smoke in Nineteenth-Century Britain', *Caverns of Night: Coal Mines in Art, Literature, and Film*, William Thesing (ed.) (Columbia, SC: University of South Carolina Press, 2000)

Darnall, Margaretta, 'The American Cemetery as Picturesque Landscape', in *Winterthur Portfolio*, 18:4 (Winter 1983)

Daumas, Maurice (ed.), *A History of Technology and Invention*, III (New York: Crown, 1969)

Delaporte, François, *Disease and Civilisation: the Cholera in Paris, 1832* (Cambridge, MA: The MIT Press, 1986)

—— *Nature's Second Kingdom: Explorations of Vegetality in the Eighteenth Century* (Cambridge, MA: The MIT Press, 1982)

Desmond, Ray, 'Who Designed the Palm House in Kew Gardens?', in *Kew Bulletin*, 27:2 (1972)

DeZurko, Edward, *Origins of Functionalist Theory* (New York: Columbia University Press, 1957)

Diestelkamp, Edward, 'The Design and Building of the Palm House, Royal Botanic Gardens, Kew', in *Journal of Garden History*, 2:3 (July–September 1982)

Donald, James, *Imagining the Modern City* (London: Athlone Press, 1991)

Doyle, Sir Arthur Conan, *The Lost World. Being an Account of the Recent Amazing Adventures of Professor George E. Challenger, Lord John Roxton, Professor Summerlee, and Mr E.D. Malone of the 'Daily Gazette'* (London, 1912)

Duckworth, Alistair, 'Fiction and Some Uses of the Country House Setting from Richardson to Scott', *Landscape in the Gardens and the Literature of Eighteenth-Century England* (Los Angeles: The William Andrews Clark Memorial Library, 1981)

Dudley, Edward and Novak, Maximillian (eds), *The Wild Man Within: an Image in Western Thought from the Renaissance to Romanticism* (Pittsburgh, PA: University of Pittsburgh Press, 1972)

Dutton, Harold, *The Patent System and Inventive Activity During the Industrial Revolution, 1750–1852* (Manchester: Manchester University Press, 1984)

Eliade, Mircea, *The Forge and the Crucible* (London: Rider, 1962)

Elias, Norbert, *Time: An Essay* (Oxford: Blackwell, 1992)

Etlin, Richard, *The Architecture of Death: the Transformation of the Cemetery in Eighteenth-Century Paris* (Cambridge, MA: The MIT Press, 1984)

Evans, Robin, 'Figures, Doors and Passages', in *Architectural Design*, 4 (April 1978)

—— *The Fabrication of Virtue: English Prison Architecture, 1750–1840* (Cambridge: Cambridge University Press, 1982)

Eyler, Ellen, *Early English Gardens and Garden Books* (Ithaca, NY: Cornell University Press, 1963)

Fisher, John, *The Origins of Garden Plants* (London: Constable, 1982)

Foster, Hal (ed.), *Vision and Visuality* (Seattle, WA: Bay Press, 1988)

Foucault, Michel, *Madness and Civilization: a History of Insanity in the Age of Reason* (New York, 1973; first published in French, 1965)

—— *The Birth of the Clinic: an Archaeology of Medical Perception* (London: Tavistock Press, 1973, ₁1963)

—— *The Order of Things: an Archaeology of the Human Sciences* (New York: Vintage Books Edition, 1973, ₁1966)

—— *Discipline and Punish: the Birth of the Prison* (New York, 1977; first published in French, 1975)

French, Stanley, 'The Cemetery as Cultural Institution: the Establishment of Mount Auburn and the "Rural Cemetery" Movement', in Stannard, David (ed.), *Death in America* (Pennsylvania: University of Pennsylvania Press, 1975)

Girouard, Mark, *Life in the English Country House* (New Haven, CT: Yale University Press, 1978)

—— *The Victorian Country House* (New Haven, CT: Yale University Press, 1979)

Gloag, John, *Mr. Loudon's England: the Life and Work of John Claudius Loudon, and his Influence on Architecture and Furniture Design* (Newcastle upon Tyne: Oriel Press, 1970)

Goodman, David, 'Fear of Circuses: Founding the National Museum of Victoria', in *Continuum*, 3:1 (1990)

Greco, Monica, 'Psychosomatic Subjects and the "Duty to be Well": Personal Agency Within Medical Rationality', in *Economy and Society*, 22: 3 (August 1993)

Grove, Richard, *Green Imperialism: Colonial Expansion, Tropical Island Edens, and the Origins of Environmentalism, 1600–1860* (Cambridge: Cambridge University Press, 1995)

Hacking, Ian, *Representing and Intervening* (Cambridge: Cambridge University Press, 1983)

—— *The Taming of Chance* (Cambridge: Cambridge University Press, 1990)

—— *Rewriting the Soul, Multiple Personality and the Sciences of Memory* (Princeton, NJ: Princeton University Press, 1995)

—— 'Memory Sciences, Memory Politics', in Antze, Paul and Lambek, Michael (eds), *Tense Past: Cultural Essays in Trauma and Memory* (London: Routledge, 1996)

Halliday, Stephen, *The Great Stink of London* (Stroud, Gloucestershire: Sutton, 1999)

Hammond, John, *The Camera Obscura* (Bristol: Hilger, 1981)

Harrison, Molly, *The Kitchen in History* (New York: Charles Scribner's Sons, 1972)

Harvey-Gibson, Robert, *Outlines of the History of Botany* (New York: Arno Press, 1981, ₁1919)

Hassan, John, 'The Growth and Impact of the British Water Industry in the Nineteenth Century', *The Economic History Review*, 38:4 (November 1985)

Hersey, George, 'J.C. Loudon and Architectural Associationism', in *Architectural Review*, 144 (August 1968)

—— *High Victorian Gothic: a Study in Associationism* (Baltimore, MD: Johns Hopkins University Press, 1972)

—— *The Monumental Impulse* (Cambridge, MA: The MIT Press, 1999)

Hirst, Paul, 'Foucault and Architecture', in *AA Files*, 26 (Autumn 1993)

—— and Woolley, Penny, *Social Relations and Human Attributes* (London: Tavistock, 1982)

—— 'Nature and Culture in Social Science: the Demarcation of Domains of Being in Eighteenth Century and Modern Discourses', in *Geoforum*, 16:2 (1985)

Hix, John, *The Glasshouse* (London: Phaidon, 1996, ₁1974)

Howard, Albert, *Darwin on Humus and the Earthworm* (London: Faber and Faber, 1945)

Howie, Will, 'Restoration of the Temperate House', in *Architects' Journal*, 175:24 (16 June 1982)

Hughes, Thomas (ed.), *Selections from 'Lives of the Engineers'* (Cambridge, MA: The MIT Press, 1966)

Hunt, John Dixon, *The Figure in the Landscape: Poetry, Painting, and Gardening During the Eighteenth Century* (Baltimore, MD: Johns Hopkins University Press, 1976)

Hunter, Ian, 'Reading Character', in *Southern Review*, 16:2 (July 1983)

—— *Culture and Government: the Emergence of Literary Education* (Basingstoke: Macmillan, 1988)

Illingworth, John, 'Ruskin and Gardening', in *Garden History*, 22:2 (1994)

Jacob, François, *The Logic of Living Systems: a History of Heredity* (London: Allen Lane, 1974, ₁1970)

Jalland, Pat, *Death in the Victorian Family* (Oxford: Oxford University Press, 1996)

Jankovic, Vladimir, *Reading the Skies: a Cultural History of English Weather, 1650–1820* (Chicago: University of Chicago Press, 2000)

Jardine, Nicholas, Secord, James and Spary, Emma (eds), *Cultures of Natural History* (Cambridge: Cambridge University Press, 1996)

Jardine, Lisa, *Ingenious Pursuits* (London: Little, Brown and Co., 1999)

Joyce, Paul, *A Guide to Abney Park Cemetery* (London, 1994)

Kern, Stephen, *The Culture of Time and Space 1880–1918* (Cambridge, MA: Harvard University Press, 1983)

Kiberd, Declan, 'Yeats, Childhood and Exile', in Hyland, Paul and Sammells, Neil (eds), *Irish Writing: Exile and Subversion* (London: Macmillan, 1991)

Kirmayer, Laurence, 'Landscapes of Memory: Trauma, Narrative, and Dissociation', in Antze, Paul and Lambek, Michael (eds), *Tense Past: Cultural Essays in Trauma and Memory* (London: Routledge, 1996)

Kohlmaier, George, *Houses of Glass: a Nineteenth Century Building Type* (Cambridge, MA: The MIT Press, 1986)

Kroeber, Karl, *Romantic Narrative Art* (Madison, WI: University of Wisconsin Press, 1960)

Lampard, Eric, 'The Urbanizing World', in Dyos, Harold and Wolff, Michael (eds), *The Victorian City: Images and Reality* (London: Routledge, 1973)

Lane, Dorothy, *The Island as Site of Resistance* (New York: Peter Lang, 1995)

Latour, Bruno, *Science in Action* (Cambridge, MA: Harvard University Press, 1987)

Lavin, Sylvia, 'The History of Ethnography', *Quatremère de Quincy and the Invention of a Modern Language of Architecture* (Cambridge, MA: The MIT Press, 1992)

Leavitt, Judith, and Numbers, Ronald (eds), *Sickness and Health in America: Readings in the History of Medicine and Public Health* (Madison, WI: Wisconsin University Press, 1978)

Le Corbusier, *Towards a New Architecture*, trans. from the French by Frederick Etchells (London: Architectural Press, 1946, $_1$1931)

Levine, George, *Darwin and the Novelists: Patterns of Science in Victorian Fiction* (Cambridge, MA: Harvard University Press, 1988)

Lovejoy, Arthur, *The Great Chain of Being: a Study of the History of an Idea* (Cambridge, MA: Harvard University Press, 1936)

Loxley, Diana, *Problematic Shores: the Literature of Islands* (London: Macmillan, 1990)

MacDougall, Elisabeth (ed.), *John Claudius Loudon and the Early Nineteenth Century in Great Britain* (Washington, DC: Dumbarton Oaks Trustees for Harvard University, 1980)

Mackenzie, John, 'Empire and the Ecological Apocalypse: the Historiography of the Imperial Environment', in Griffiths, Tom and Robin, Libby (eds), *Ecology and Empire: Environmental History of Settler Societies* (Melbourne: Melbourne University Press, 1997)

Malin, Nadav, 'Indoor Air Quality: Some Old and Some New Sick-Building Culprits', in *Architectural Record*, 184:2 (February 1996)

Mallgrave, Harry, *Gottfried Semper: Architect of the Nineteenth Century* (New Haven, CT: Yale University Press, 1996)

McCracken, Donal, *Gardens of Empire: Botanical Institutions of the Victorian British Empire* (London: Leicester University Press, 1997)

Meller, Hugh, *London Cemeteries, an Illustrated Guide and Gazetteer* (Amersham: Avesbury Publishing Comany, 1985)

Minter, Sue, *The Greatest Glass House* (London: HMSO, 1990)

Morgan, Joan and Richards, Alison, *A Paradise Out of a Common Field* (New York: Harper and Row, 1990)

Morton, Alan, *History of Botanical Science: an Account of the Development of Botany from Ancient Times to the Present Day* (London: Academic Press, 1981)

Murphy, Shirley, *Our Homes and How to Make them Healthy* (London: Cassell, 1885)

Nead, Lynda, *Victorian Babylon: People, Streets, and Images in Nineteenth-Century London* (New Haven, CT: Yale University Press, 2000)

Oelschlaeger, Max, *The Idea of Wilderness: from Prehistory to the Age of Ecology* (New Haven, CT: Yale University Press, 1991)

Perez-Gomez, Alberto, *Architecture and the Crises of Modern Science* (Cambridge, MA: The MIT Press, 1983)

Pevsner, Nikolaus, *Pioneers of Modern Design* (Harmondsworth: Penguin, 1975, $_1$1936)

Prest, John, *The Garden of Eden: the Botanic Garden and the Re-Creation of Paradise* (New Haven, CT: Yale University Press, 1981)

Ray, Isaac, *Mental Hygiene*, with a new introduction by Frank Curran (New York and London: Hafner Publishing Company, 1968)

Rose, Nikolas, *The Psychological Complex: Psychology, Politics, and Society in England, 1869–1939* (London: Routledge, 1985)

Rose, Hilary and Steven (eds), *Alas, Poor Darwin: Arguments Against Evolutionary Psychology* (London: Vintage, 2001)

Rostron, Jack, *Sick Building Syndrome: Concepts, Issues and Practice* (London: Spon, 1997)

Rowe, D. M., 'Sick Building Syndrome: the Mystery and the Reality', in *Architectural Science Review*, 37 (September 1994)

Russell, Bertrand, *A History of Western Philosophy* (London: Unwin Hyman, 1984)

Russell, Colin, *Science and Social Change* (London: Macmillan, 1983)

Rykwert, Joseph, *On Adam's House in Paradise* (New York: The Museum of Modern Art, 1972)

Shaftel, Norman, 'A History of the Purification of Milk in New York, or, "How Now, Brown Cow"', in Leavitt, Judith and Numbers, Ronald (eds), *Sickness and Health in America: Readings in the History of Medicine and Public Health* (Madison, WI: Wisconsin University Press, 1978)

Shattuck, Roger, *The Forbidden Experiment: the Story of the Wild Boy of Aveyron* (New York: Farrar, Straus and Giroux, 1980)

Shteir, Ann, *Cultivating Women, Cultivating Science: Flora's Daughters and Botany in England, 1760–1860* (Baltimore, MD: The Johns Hopkins University Press, 1996)

Simo, Melanie Louise, *Loudon and the Landscape* (New Haven, CT: Yale University Press, 1988)

Singer, Charles (ed.), *A History of Technology*, V (Oxford: Oxford University Press, 1954–84)

Sproule, Ann, *The Social Calendar* (Poole: Blandford Press, 1978)

Stafford, Barbara, 'Images of Ambiguity', in Miller, David and Reill, Peter (eds), *Visions of Empire: Voyages, Botany and Representations of Nature* (Cambridge: Cambridge University Press, 1996)

Stark, Werner, *Jeremy Bentham's Economic Writings*, I (London: Allen and Unwin, 1952)

Stenick, Nicholas, *Science and Creation in the Middle Ages* (Notre Dame, IN: University of Notre Dame Press, 1976)

Summerson, John, *Victorian Architecture: Four Studies in Evaluation* (New York: Columbia University Press, 1970)

Tate, Alan, 'Loudon and the Return to Formality', in MacDougall, Elisabeth (ed.), *John Claudius Loudon and the Early Nineteenth Century in Great Britain* (Washington, DC: Dumbarton Oaks Trustees for Harvard University, 1980)

Thesing, William, *Caverns of Night: Coal Mines in Art, Literature, and Film* (Columbia, SC: University of South Carolina Press, 2000)

Tristam, Philippa, *Living Space in Fact and Fiction* (London: Routledge, 1989)

Vesely, Dalibor, 'Architecture and the Conflict of Representation', in *AA Files*, 8 (January 1985)

Vidler, Anthony, 'Rebuilding the Primitive Hut', in *The Writing of the Walls: Architectural Theory in the Late Enlightenment* (Princeton, NJ: Princeton Architectural Press, 1987)

—— 'Psychopathologies of Modern Space: Metropolitan Fear from Agoraphobia to Estrangement', in Roth, Michael (ed.), *Rediscovering History: Culture, Politics, and the Psyche* (Stanford, CA: Stanford University Press, 1994)

Warner, George, *Landmarks in English Industrial History* (London: Blackie and Son, 1924, ₁1899)

Waters, Michael, *The Garden in Victorian Literature* (Aldershot: Scolar Press, 1988)

Whorton, James, '"Tempest in a Flesh-Pot": the Formulation of a Physiological Rationale for Vegetarianism', in Leavitt, Judith and Numbers, Ronald (eds),

Sickness and Health in America: Readings in the History of Medicine and Public Health (Madison, WI: Wisconsin University Press, 1978)

Wittkower, Rudolf, *Architectural Principles in the Age of Humanism*, 4th edn (London: Academy Editions, 1973, ₁1949)

Woods, Mary and Warren, Arete, *Glasshouses* (New York: Rizzoli, 1988)

Young, Robert, *Mind, Brain and Adaptation in the Nineteenth Century* (Oxford: Clarendon Press, 1970)

Index

Illustrations are indicated by italicized numbers